Applied Agricultural Research

Applied Agricultural Research

Foundations and Methodology

Chris O. Andrew
and Peter E. Hildebrand

Routledge
Taylor & Francis Group
LONDON AND NEW YORK

First published 1993 by Westview Press

Published 2018 by Routledge
52 Vanderbilt Avenue, New York, NY 10017
2 Park Square, Milton Park, Abingdon, Oxon OX14 4RN

Routledge is an imprint of the Taylor & Francis Group, an informa business

Library of Congress Cataloging-in-Publication Data
Andrew, Chris O.
Applied agricultural research : foundations and methodology / by Chris O. Andrew and Peter E. Hildebrand.
p. cm.
Includes bibliographical references (p.) and index.
ISBN 0-8133-8781-7
1. Agriculture—Research. 2. Agriculture—Research—Methodology.
I. Hildebrand, Peter E. II. Title.
S540.A2A52 1993
630'.72—dc20 93-31763
CIP

ISBN 13: 978-0-367-01191-8 (hbk)
ISBN 13: 978-0-367-16178-1 (pbk)

Contents

PART TWO
PLANNING APPLIED RESEARCH

PART THREE
CONDUCTING APPLIED RESEARCH

Figures

Preface

This book is based on *Planning and Conducting Applied Agricultural Research* by the same authors. Revisions of all of the chapters and inclusion of four new chapters prompted a new title. The first book (published in Spanish by ICTA, Guatemala, 1977; in English by MSS Information Corporation, New York, 1976; and in English by Westview Press, Boulder, 1982) was the culmination of a group effort to eliminate a deficiency made evident during the organization of a graduate course in research methodology at the UN-ICA Graduate School of Agricultural Sciences in Bogota, Colombia.[1] The deficiency involved the difficulties in organizing research projects oriented toward real world problems and formulated so that (1) the research could be completed within the available period of time, and (2) the results would be useful in helping to resolve the problem toward which the study was directed. It became obvious to the agricultural economists[2] working with the graduate program that many approaches to the presentation of research methodology were not successful in helping students become efficient researchers consistently able to make meaningful contributions to the resolution of agricultural and related problems of their country.

---

[1]The graduate school was jointly administered by the National University of Colombia (UN) and the Colombian Agricultural Institute (ICA). Besides the graduate school, ICA held responsibilities in research and extension as well as service activities such as control of agricultural chemicals and port sanitation.

[2]The group consisted of several Colombian agricultural economists with ICA and others with the University of Nebraska Mission in Colombia. This technical assistance team worked with ICA, National University and the Graduate School from 1967 to 1972.

Initially the challenge was focused upon new means of presenting the requisites for successful research in a manner that would have real meaning and utility for the students. This germinated the core idea presented in Chapter 6. These topics are not new; most research methodology material discusses research problems, hypotheses, and objectives. It is the means of fitting these parts together into a more useful form which is important. During several courses and while counseling students and fellow staff members of ICA in research, we were able to revise and improve the concepts. With the new book comes experience in many courses and comments by various instructors and students who have used the material over twenty years. We have found that this approach to planning and executing applied research can be experienced, understood and used. This has prompted revisions to the original text and additions to strengthen capacity to plan and execute applied research.

In developing the book we found that although it was relatively easy to use the approach in training students and counseling researchers, it was difficult to present the approach in a form understandable and usable by persons with whom we would have no direct contact. The inclusion of project proposals is designed to help overcome this concern; for the instructor or course leader, a series of experiential activities and exercises are recommended to assist in developing and managing a learning environment for students of applied research.

Although many of the examples are authentic, and some are drawn from Colombia where most of the first book was written, the research difficulties presented are not unique to any country; they are encountered throughout the world, including the United States. Our desire is to provide researchers, wherever they may be, with an approach to research under various time and resource limitations which will help them be of greater service to their clients. This is particularly important in developing countries where applied research is so needed; it is also important in more advanced countries where research resources are increasingly restricted.

Chris O. Andrew
Peter E. Hildebrand

Acknowledgments

It is not possible to acknowledge all the ideas and assistance we have experienced from our generous students and peers over the twenty-five years we have worked on the topic of applied research methodology. Ideas and most of the early drafts for the first book were forthcoming while the authors were with the University of Nebraska and assigned to the Colombian Agricultural Institute and the National University of Colombia. To our colleagues, and to students at these institutions, we offer sincere acknowledgments and hope that this book will be useful to them. We appreciate the many constructive comments by students directed to improving the learning potential of the book.

Also, we extend our appreciation to the Ministry of Agriculture of El Salvador, where the second author was stationed in the early 1970s on a technical assistance contract, for translation and preliminary publication of the first manuscript in Spanish. For final reviews of the Spanish translation and preparation of that manuscript for printing, gratitude is extended to the Guatemalan Institute of Agricultural Science and Technology where the second author was assigned in the late 1970s while employed by the Rockefeller Foundation.

We express our appreciation for assistance received from the Food and Resource Economics Department at the University of Florida. To the late Fred Prochaska and to his students who used the first text for fifteen years in the research methodology course, we are grateful for constructive criticism. Similarly, new materials in this book have been tested with numerous students in the research methodology and farming systems courses taught by the authors at the University of Florida. Special recognition is due Leo Polopolus, chairman of the Food and Resource Economics Department (1973-1983), and the late W.W. McPherson, graduate research professor, for reviews and consultation. Similarly, the encouragement to prepare the new text and willingness to grant time for that purpose by current Chairman Larry Libby was important. Appreciation is extended to Sharon Bullivant in the Food and Resource Economics Word Processing unit for camera-ready preparation of this text and to RoseMary Espaillat for preparation of the figures.

C.O.A.
P.E.H.

1

Introduction

Lingering on and ever more demanding is the desire to resolve problems through pragmatic observation, systematic evaluation and concerted action. The process repeats itself. Each new generation of research scientists emerges to confront the challenge of food production and distribution, to face extensive world hunger and malnutrition. Additionally the natural and human resource bases for agricultural production, as well as the research resource base for technology development to enhance productivity, are becoming more difficult to sustain. But the demand for effective and efficient research methodology directed to resolving real-world problems continues and increasingly is more important.

As an age-old task that bears individual, social and sometimes cultural characteristics, problem resolution is essentially an experiential process. Routine and methodology can enhance both the process and the experience. In tackling the development of a course on research methodology more than twenty-five years ago and in preparation of the text *Planning and Conducting Applied Agricultural Research*, we aspired to teach young agricultural economists to become researchers. We wanted to improve the process of problem identification, of selecting alternatives and priorities to help explain and resolve the problem, and the specification of research objectives to effectively and efficiently guide the needed tasks of observation and analysis. Applying methodology to problem solving, much as with reading or speaking, has natural tendencies. Proficiency and efficiency, however, in research require accumulation and transfer of individual and communal experience. But if we accept that the approach to research is a relatively deliberate yet natural progression to finding solutions, and note that method and technique often dominate present disciplinary thought and application, what role is there for instruction in research methodology?

The questions are: "Can the process of research be taught?" and "Will research technique and discipline bow to the broader process of research?" We believe that little can be taught but that a great deal can be learned in an enhanced experiential context. In its deceiving simplicity, this text attempts to help structure a learning environment for improving skills and processes in the methodology of applied research. Problem resolution displays cultural characteristics and individual initiative of each person in a cultural group. Scientific research intensifies this quest for knowledge to resolve problems. Learning and conditioning the methodology of research continues as each new generation of researchers enters the subculture of scientific investigators.

Misidentification of researchable problems, selection of non-priority alternative solutions, and use of inappropriate tools of analysis persist. In the face of evermore complex concerns and resource limitations, there remains ample room for becoming more efficient and effective in service to humankind. We humbly remind you that the young man identified in our first book is somewhere out there still seeking to understand the problem of his client...

About midmorning on a fine September day, the Minister of Agriculture is just completing a phone conversation as the door to her office opens to admit one of the young men in the research group from the Planning Department of the Ministry. At about the same time the Vice Minister and the head of planning enter and the four people seat themselves around a small conference table. The Minister opens the meeting, directing her comments to the Vice Minister: "I'm told that we have a final report on that project on importing fertilizer, that's great. It's just in time. Now we ought to be able to convince the import-export people that they can't reduce our request for fertilizer and still expect us to meet our new trade commitments."

Turning to the young researcher, the Minister says encouragingly, "OK, young man, they tell me you've made an excellent study. Now tell us exactly how much fertilizer is it going to take over the next five years to meet our production targets?"

"Ma'am?" replies the researcher, apparently a little confused.

"Come, come, now, don't be nervous. Just tell us about the research you've been doing. What are the results?" The Minister realizes the researcher may be a bit timid.

"Oh, yes, Ma'am," smiles the researcher, "thank you."

"Well, as soon as we got your request for information on the importance of fertilizer, we checked to see what data we had on Japan and some other countries where there have been recent increases in production. We thought this ought to give us some good ideas about the relationship between fertilizer use and crop production. Here, Ma'am, we have a series of graphs showing the correlation between these two variables for a number of countries."

"Yes," replies the Minister, "it's quite evident that fertilizer is important in increasing crop output. Now, how much are we going to need?"

The researcher continues, "According to the latest census, which unfortunately is several years old as you know, only about 40 percent of our farmers are using any fertilizer. This is considerably below the rate in the other countries I mentioned. And in those countries income per farm family has been increasing rapidly, again demonstrating the importance of fertilizer." The minister nearly interrupts but lets the researcher continue. "Now if we want to double the number of farmers using fertilizer, we might be able to assume that we need twice as much fertilizer as now."

"Yes, I suppose," replies the Minister, "but what about the land area involved and what about the requirements for the different crops?"

The researcher, thumbing back through his report answers, "We don't have any information on area of each crop that is fertilized but cotton and sugarcane consume about 80 percent of the fertilizer used and..."

He is interrupted by the Minister who says, "But don't you have any estimates of the quantities required to get the production we need over the next five years? Our problem is that they want to cut back on fertilizer imports to help domestic fertilizer production just at the time we have to try to increase crop production in a big hurry. We need to know how that could affect our program and how much importation we need to ask for." Turning to the head of the planning group she says, "I thought we discussed this pretty thoroughly that day it came up. What happened?"

"Well, that's right, and I knew we were going to have a hard time convincing them how important fertilizer was in our program and I know we talked about that in my office, didn't we?" the head of planning asks the young researcher.

"Yes, sir," replies the young man, "we knew we had to get you some good information on the importance of fertilizer and that's why I have this information on Japan and those other countries."

> The most serious consequence of weak applied research is the cost associated with not having relevant information for making decisions.

"Well," replies the Minister, "that's not going to be much help in solving our problem. We have the meeting with them tomorrow. But maybe we could hold off a decision for a couple of days if you think you can get me the information we need by that time. Why don't you call someone at the experiment station? Maybe they can help you. But you better get going. Don't forget how important fertilizer is to us..."

There are several important points in this dialogue which, though fictional, represents a real-life situation encountered much too frequently. The most important point is that after waiting right up to the deadline, the Minister, who is the client, did not obtain the information she needed for a meeting of great importance. As a result, the research costs incurred by the Planning Department for this project yielded little of value. The most serious consequence, of course, is the cost associated with not having the relevant information for making the decision at the meeting. Although some of the reasons for the unfortunate situation are evident in the dialogue, others are more subtle. Regardless of the reasons, we hope this book will contribute to the more effective use of research resources and help prevent the kind of unhappy discussion as that between the Minister and the researcher.

Applied Research

Research is the orderly procedure by which people increase their knowledge and is contrasted to accidental discovery because it follows a series of steps designed precisely for the purpose of developing information.[1] Knowledge gained by research may be used by people to produce a greater abundance of food and fiber, to lighten the burdens of their labor, or in any number of ways to generally improve their well being. Or new knowledge may simply be added to the store of concepts about the universe to await application at some future point in time. Research undertaken specifically for the purpose of obtaining information to help resolve a particular problem is applied research. For a research undertaking to be applied research it is not necessary that the results (the new knowledge) resolve or help resolve the problem which initiated the project (though hopefully they will), but it is necessary that the research have a specific problem orientation. It is this kind of research, oriented to resolution of specific problems, toward which this book is directed.

The development of Mexican or dwarf wheat was the result of an applied research process oriented toward the resolution of a specific problem [Paarlberg, 1969]. A fertilizer experiment oriented toward making recommendations to farmers is another example, as is the work of a government planner trying to estimate the likely supply response to fertilizer for a particular commodity under a proposed new program, or the total fertilizer requirements necessary to reach specified production goals, as in our dialogue. Determining acceptability of a newly developed feed concentrate for fattening hogs in tropical areas or the development of a hand seeder for steep terrain in primitive areas, assuming the lack of each represented a problem, also would be classified as applied research. In general, the research referred to in this book is oriented toward providing useful information to decision makers such as farmers and public administrators.

> An effective applied research methodology is directed toward the efficient use of available research resources to maximize the probability of achieving meaningful results to help resolve problems.

[1]For thoughts concerning the meaning of research and the scientific method see Cohen and Nagel; Conant (1951a & 1951b); Dewey; Friedman; Gibson; Hildreth and Castle; Johnson; Kaplan; Miller; Murphy; Nagel; Noltingk; Pearson; Runes; Zetterberg; Back; Halter and Jack; Paarlberg (1963); Parsons; Peirce; Sundquist; Vining; Weaver.

Applied research, such as that just described, is carried out in all parts of the world. It is a much more widespread activity than fundamental research which is a necessity but one that only the wealthiest countries can afford. Most applied research is conducted under moderate to severe resource limitations which necessitate efficiency in the research process. Disappointment in the results of applied research -- a "So what?" response -- in most instances can be traced directly to the use of an inadequate and/or ineffective applied research methodology which failed to correctly identify the problem.

An effective applied research methodology is directed toward the efficient use of available research resources to maximize the probability of achieving meaningful results to help resolve problems. Recent methodological work has enhanced the productivity of research resources. Farming Systems Research and Extension (FSRE) methodology has emerged in part to help deal with limited resources for production and farm management research. Advances in interview processes have improved problem identification and analysis in marketing systems research, resulting in more effective and efficient research resource use. New approaches are emerging for the study of natural resource use and for policy formulation.

The basic theme of this book is that of applied research as a service to a client with a problem for which the information obtained by research can help resolve. Because applied research has a definite purpose, there is usually a time constraint or deadline within which the work must be completed as well as a limit on the other resources the client has available or is willing to use in the resolution of the particular problem. Consequently, the researcher must be cognizant of the efficient use of the research resources while at the same time functioning so as to maximize the likelihood of providing a useful product to the client. The book is also useful to graduate students who are trying to define and execute a project that is subject to completion in a specified (finite) time period.

Besides the approach to applied research (covered in the book), another important factor affecting success in serving clients is the research environment within which researchers labor. An applied researcher cannot be effective in satisfying clients when he or she is isolated from them by a system that reduces or prevents effective communication.[2] This can happen, for example, when extension service personnel with direct client contact have little communication with researchers even though they may be in the same organization. It can also happen in research organizations in which projects are dictated by administrators who have little contact with the clients and hence have no appreciation of their real problems. This could be an argument for maintaining small research organizations. More importantly, it is an argument for an organization in which researchers maintain close personal contact with their clients and where researchers in turn share in determining the research priorities of the organization. Recent experience with on-farm research has demonstrated the value of research, extension and farmer (client) collaboration. We suggest that a better coordinated working relationship between research and extension administrators, researchers, extension people and clients will develop if all understand the approach to problem identification which is presented in this book.[3]

The Book

Perhaps the most critical deficiency in methodology is the failure to adequately identify the specific problem toward which research is to be oriented (as happened to the researcher in the Ministry). This may result when the researcher uncritically accepts the problem as stated by the client or by his or her spokesperson (the "importance of fertilizer" was not the Minister's problem). It may also happen when one is not really sure who the client is -- farmer, policy maker, marketing firm, consumer, peer, administrator or someone else. Another serious deficiency may occur even after properly identifying the problem. This is the failure to formulate hypotheses and objectives correctly oriented toward the resolution of the problem or failure to use appropriate

---

[2]See Ackroff; Schuh; Ayer and Schuh; Kiehl; Knoblauch, Low and Myer; McKee; Motes; Plaxico; Stern; Thirring; White; Wilson; Zimet; and McDermott and Andrew for readings on the role and impact of institutions in the research process.

[3]With reference to administration and management of research and extension institutions, note the companion text prepared under the principle authorship of J. Kenneth McDermott.

analytical techniques and theory (what hypotheses did the Ministry researcher use?). The most critical components of the process are formed from the interrelationships exposed by problem identification, hypotheses, objectives, analytical techniques, and resource constraints. These topics are all covered in the book.

Practical experience is invaluable in helping the researcher overcome the obstacles which are so often encountered in applied research. In all phases of the material to be presented, the difficulties associated with sub-optimum research conditions under which the individual researcher is apt to be working are considered.

In this book, we have divided the topics into separate chapters and the chapters into three parts. Neither the parts nor the materials in any chapter are independent.

In Part One, we address the role of theory and *your* (the researcher's) role in research, two very important, but often overlooked, resources. We are concerned with the necessary linkage between your problem solving skills, in the process of knowledge gathering and application, and the process of structuring knowledge commonly found in the development of theory.[4]

> Perhaps the most critical deficiency in methodology is the failure to identify the specific problem.

Planning activities are discussed in Part Two of the book. In planning the research project, one must always assess the means available for conducting the research, and during the research process it may be necessary to modify portions of the original plan. Each of the activities is affected by the others and by the research resource constraints under which the researcher is toiling. Effective problem identification is essential. The kinds and sources of data to be used and the methods of analysis will be dictated by the hypotheses and objectives, but they, in turn, must be finalized only after having taken into account the effect of resources on availability of data and/or analytical competence. In Part Three, Conducting Applied Research, we discuss data collection, verification, interpretation, and presentation of the results to the client.

---

[4]Readers who are more exclusively concerned with planning and conducting applied research, and less with foundations for it, may prefer to begin with Part Two.

An appendix provides a cross section of applied research project proposals to use as cases for further study. Each could be improved by continued revision based on the principles presented in the book.

Because of the deliberate and intense emphasis placed upon problem identification, the book does not discuss in detail each basic element of a research project. One should be cognizant, however, of the interrelated components of a research project and employ them when planning and conducting applied research. These essential elements are: (1) a problem statement accompanied by sufficient information to justify the need for research; (2) hypotheses; (3) objectives; (4) budget; (5) the appropriate theoretical and analytical approach and procedures; (6) data requirements including sources and procedures for obtaining data; (7) a detailed work plan showing jobs to be done and time sequences; and (8) the reports to be issued for each audience.

PART ONE

Foundations and Applied Research

2

Creative Applied Research and You, the Researcher

You are the most important research resource you will ever manage. You could become an award winning research scientist, problem solver, educator, manager or leader. You alone can develop and draw from your full research capability. You are challenged to conduct research for creative and useful problem solving. You are called on to apply both an effective and an efficient methodology. Finally, you are further challenged to carefully consider the ethical implications of your research and results.

Management of research begins and ends with management of one's self as the primary research resource. In that capacity one manages one's self to gain analytical and conceptual creativity.[1] Of course time and other resource management issues are important, but in the end, effective management of those resources remains a problem of managing one's own capabilities. These capabilities extend from the conscious to the subconscious and bear ultimately on the results. Thus, applied research management is management of self and of a learning environment that is directed to problem solving ends.

> Research begins and ends with management of one's creative self.

---

[1]Recall how Boulding in opening his book, *The Image*, is creating and interacting with his environment as a means for conveying knowledge gained by observation, by research.

Enriching the Research/Learning Environment

Usually we judge performance from an orderly, linear or vertical thought perspective that suggests logical sequences and well ordered results. This is as it should be. Yet creative solutions often result from disorder, random occurrence and a lateral thought perspective. Here a distinction is made between vertical and lateral thought, not to exclude either from the tools of the researcher, but to enrich the learning and discovery process. The goal is to reach the highest human potential for effective management of research resources to achieve useful results. Most of this chapter deals with lateral modes as a complement to the vertical modes evident in much of the rest of the book. Further insights into lateral thinking are available in de Bono.

Our focus is on how the researcher thinks or manages thought. Thinking for learning or research purposes may be classified as insightful, sequential and strategic. *Insights* result when solutions seem to appear from a sudden jump in thinking; the steps leading to the solution are not apparent. *Sequential* thinking, to the contrary, follows a solution path in a progressive sequence of activities based in modification, trial and error, improvement, new ideas and mistakes. The sequence is logical, with identifiable steps or activities occurring one after another (as an example, see Chapter 5 in its entirety). *Strategic* thinking does not seek a definite solution so much as a pattern of behavior that will be more effective than others; it selects the most appropriate steps or activities for problem solving from many alternatives. Both vertical and lateral thought processes contribute to insightful, sequential and strategic thinking.

One may be tempted to identify the realist view point (discussed in Chapter 3) with lateral thought and the instrumentalist perspective with vertical thought. However, both vertical and lateral thought do unavoidably serve both realists and instrumentalists.

Balancing Vertical and Lateral Thought

The following lists contrast vertical and lateral thinking as a basis for introducing the potential for optimal "invention" or problem solving:

Vertical	Lateral
• Analytical and logical	• Insightful, provocative and arbitrary
• Respectable and rigid	• Suspicious and rationalizing
• High probability/low risk	• Low probability/high risk
• Elaborative yet selective	• Random yet inclusive and generative
• Smooth, organized and sequential	• Ragged, chaotic and can make jumps
• Uses chaos through absence of direction (relegated to error)	• Uses chaos by direction
• Logic is in control of the mind-conditioning	• Logic is at the service of the mind -- discovery and rediscovery
• Moves only if there is a direction in which to move	• Moves in order to generate a direction
• Correct at every step (required)	• May skip steps or be incorrect
• Concentrates and excludes irrelevant	• Welcomes chance intrusions
• Labels, classifications and categories are fixed	• Labels, classifications and categories are fluid
• Follows the most likely paths	• Explores the least likely paths
• A finite process	• A probabilistic process
• Uses negative to block pathways	• Are no negatives

Some general principles of lateral thought further suggest ways for mobilizing lateral processes as a complement to the more conventionally accepted vertical processes:

- Recognize dominant or polar ideas -- deliberately pick and define dominant ideas, and gradually distort each idea until it loses identity. Insights from the interactive process of removing identity of a polar idea may yield major questions that lead to alternative research hypotheses, new conceptualizations or remove and challenge a conventional wisdom.
- Search for different ways of viewing things -- is the glass half empty or half full? Does it make a difference? In what way?
- Relax the rigid control imposed by vertical thinking -- permit chaos to enter and reduce "fixed direction" in the thinking process, skip some "logical" steps to see what could emerge.
- Use chance -- expect and learn from the unpredictable, brain storm; permit thought contamination (read widely), use serendipity.
- Use the memory surface provided by the subconscious mind -- the memory surface provides a basis upon which information can be self organizing. This is necessary because the conscious mind has a short attention span. A short attention span is good for in-depth focus while the memory surface provides a basis for finding ideas for addressing new situations.

> A mind prepared for research draws on conscious and sub-conscious potentials.

The balance of lateral and vertical thought activity applied to problem solving or "invention" includes at least five interactive stages. A person works typically in two or more stages but does not do so always in a sequence. Parts of the "invention" may pass through the stages separately while at other times as part of the whole. Illumination may come first from subconscious mental activity and thereby demand preparatory work which will initiate the process. The stages include:

- Preparation -- primarily vertical and conscious mind processes: traditional logic, math forms and experimental science, repeated systematic examination, refinement.
- Imagination -- primarily lateral and conscious to subconscious processes: alternate views, new descriptions, creative thought, open, expansive, synthesis to synopsis (to see things that never were).

- Incubation -- subconscious, involuntary mental activity conditioned by an open and free environment to nurture the subconscious mind.
- Illumination -- sudden enlightenment along with recording, structuring and strengthening the results.
- Verification -- tests of illumination against practice and/or logic as a means for validation and redefinition of the need for further inventive activity.

YOU Management

Why manage yourself? Conscious, rational, objective activity is but a narrow potential of individual capability. It is limited by conscious processes and tools. Versatility can be gained by expanding the tool kit to include not only lateral thinking with vertical thinking at the conscious level but also understanding what the subconscious can offer. Freud compares the human mind to an iceberg with one eighth of the mass above water -- only a small part of our mental processes are conscious.

George Ladd helps us with several suggestions for exploiting ourselves as research resources. The purpose is to stimulate a prepared mind for research as a complement to our conscious, objective and rational research tools. Further adaptation of Ladd's work results in the following list:

1. *Condition one's own research environment* by assuming an attitude of positive possibility -- "It can be done," "Let's try," "Yes," "We can," "There is a way," "We have the ability."
2. *Release mental locks* that inhibit you -- "The right answer," "That's not logical," "Follow the rules," "Be practical," "Avoid ambiguity," "To err is wrong," "Play is frivolous," "That's not my area," "Don't be foolish," "I'm not creative."
3. *Doubt* what is given as conventional wisdom.
4. *Allow a venturesome attitude* to confront ego, self esteem, fear of failure and apprehension of an encounter with risky projects.

5. *Provide for diverse experiences, memories and interests* as a means of providing new input and a relaxed stimulus to the subconscious.
6. *Thoroughly prepare* (in the conscious mind) by carefully formulating the problem so that causal conditions become a focus. Then use both vertical and lateral thought processes to reach a more comprehensive identification.
7. *Use tension* as a basis for conscious absorption of the problem and stimulation of the subconscious. Curiosity and frustration (cognitive dissonance) collaborate to effectively establish positive tension.
8. *Temporarily abandon* the project by going away from the problem and returning later as a process of incubation works between the conscious and subconscious mind. Alternating from conscious thought to positive abandonment can allow new insights and breaking patterns of new thought.
9. *Relax* following intense thought as a basis for favorable intuition. Find a comfortable and effective balance in managing your time -- relaxation provides optimal support to focused concentration.
10. *Write* periodically to force synthesis even when the ideas do not seem to be there. Writing cultivates an environment where the conscious mind has planted a seed and the subconscious mind may produce an insight.
11. *Record and analyze dreams* as a means for incorporating your mind's night work into your conscious research efforts [Taylor]. Often it is the more absurd (seemingly in conscious thought) dreams that yield the most insight when interpreted and often group work draws out the full content.
12. *Exchange thoughts and ideas* with colleagues to gain new dimensions while nurturing the emerging results of your own conscious and subconscious search.
13. *Free yourself from distraction* periodically to permit insight and interpretation from the subconscious. Give yourself a total break, even a few minutes here and there so that you can completely follow a thought or a day dream.[2]

---

[2]Ladd gives an example of "emphasis" from one person's view of the power found in the subconscious.

14. *Impose deadlines* to force the subconscious to work.
15. *Capture intuitions*, dreams, hunches, urges, ideas and impulses in ways that help you focus on what the subconscious product is telling you. Be alert to techniques (your own or those of others) for retaining these thoughts that appear at the edge of the conscious mind.
16. *Reunite* the lateral thinking and creative processes continually with vertical thinking processes as a means for bringing the research effort into a useful form. The word is balance.

Contrary to the intent of this chapter and certainly the principles of lateral thought, the above list would appear to have order. It should not. Each tool can be used separately and randomly yet some may find a particular order purposeful to serve their unique learning environment.

Capturing Creativity

We make a case for the open research environment. You may be stimulated to new formulations, methodological insights, interpretations and general creativity by your YOU management skills. You have a major resource in both the realms of the conscious and subconscious. Interactive exchange and complementarity can produce excellent team results or simply serve to help you advance your theory development work (knowledge structuring) and your problem solving skills (knowledge retrieval and application).

Sustainability in agriculture and resource use is a concern that speaks to a need for crossing disciplines, borrowing insights, creating new research ventures and countless realignments that stress individual researchers and research institutions. Success in these ventures may lie in a mix of vertical and lateral thought methods directed toward new methodologies and radically different theories from those which presently direct our labors. It seems that special management skills and orientations are needed to tap this potential. Is this not true, however, in every case?

Thus, one is given the task of managing the human input for the purpose of inquiry and problem resolution. One manages self and others who assist with the task. To do so requires attention to a creative process that is open and receptive yet progressive toward the purpose to which the specific inquiry is directed: a well defined problem. Thus, you are exhorted to:

> Managing creative research involves diversity, complementarity, synthesis and consensus.

1. Nurture diversity -- other alternative positions can strengthen your own by forcing improved focus, by borrowing something from the alternative and by giving another vantage point. The sum has a potential to become greater than the parts only if there are diverse elements.
2. Encourage complementarity -- finding common ground allows for the whole to become greater than the sum of the parts. The scope of solutions to well defined problems is often expanded by identifying common or shared opportunities.
3. Support synthesis -- building on common ground allows for emergence of the "hybrid" solution that may not have occurred to anyone until the integration process began.
4. Evaluate consensus -- finding direction and rejection in experience and application provides the test of our work and gives meaning to future attempts. From here we are driven again to synthesis, complementarity and diversity.

Right Process and Right Product

The ethics of the practice, the content and the results of your research are challenged by new awareness of the role humans play in altering nature. As a member of the scientific community you have a responsibility to humankind and nature to be vigilant in your reflections about the potential uses and misuses of knowledge. This further obligates you to an active concern for human values in the culture as a whole where you and we practice research. Various socio-cultural institutions and mass communication media provide some of the mechanisms for confronting those obligations. The obligation extends from a need to maintain pluralism at the society, institution and activity levels where moral concerns circumscribe the quest for knowledge and the use of that knowledge. As a researcher you have a key role to play in the formal knowledge quest by observing the appropriate moral conditions of your research environment. Yet the desire is not to constrain completely the research process. Freedom to do research must be distinguished sufficiently from the use of research so that the knowledge quest itself is preserved and sustained for future generations.

> You have an obligation to understand the moral and ethical dimensions of your research.

Appropriate use of knowledge becomes a major concern today in the areas of population and nutrition, biomedical research, sustainability in food production, and economic integration to mention a few. Approaches to research more concerned with a sustainable agriculture are being developed. Various types of multidisciplinary systems-oriented modeling activities extend from crop models through attempts at whole farm models. Here the client/subject ethical question is initially reduced as numerous situations are simulated.

Farming systems research is a case where farmers directly enter the research process. Multidisciplinary teams of biophysical and socioeconomic scientists, along with farmers, address a wide range of environmental, household and farm conditions that influence agriculture and use of natural resources. Farmer managed, on-farm research is the "scientific" mode. Experimental trials are designed with farmer participation and implementation. This keeps the practices and technologies within the socioeconomic and biophysical potential of a selected research domain. Thus, the practical and ethical questions associated with research drawn from a more "scientific" approach on distant experiment stations and imposed on the farm either directly or through "autonomous" processes, are reduced. Some basic science from those experiment stations, however, is necessary to support the on-farm research process.

Major ethical challenges to the farming systems approach are numerous.

1. Does the process of looking within for technological advances relegate subsistence farms to continuous low level subsistence?
2. Is the common good of the community and family served by the individual participating farmer? All cannot participate in the early research stages due to the great number of farms.
3. Are the multidisciplinary team members sufficiently aware of socio-cultural and religious traditions to anticipate when the process may result in coercion? What are the consequences in the process?

4. Is the natural process of adaptation to malnutrition (or starvation) more conducive to natural order than any attempt to induce technological change?
5. Does it matter that technology emerges from relatively homogeneous parts of similar farms (wholes) within a broader set of farming systems? Is the selection process discrimnatory?
6. Are equity questions of gender placing utilitarian and anthropocentric considerations above the common good?
7. How are women farmers dealt with in terms of the whole? What is the whole? And do children or the elderly fit into the picture?

Principles specifically designed to address questions of this type have not been systematically developed and tested, if considered at all. Critical to the ethical test would be to understand the research setting and give those who live in the community collective and self determining roles in the change process. While an excellent example of an ethical challenge to the research process, farming systems research is but one of many approaches involving clients directly in the technology innovation process.

The ethical conscience must test principles within the particular research context or setting as a basis for moral intelligence. Your responsibility as a researcher is indeed quite extensive, yet you can proceed. Do responsible research. Know that use of your results requires yet another set of ethical considerations. In the first case we must know so that in the second case we have something to make "wise" and "ethical" decisions about. We believe that research is a very necessary task that can not avoid questions of ethics. YOU are essential to that task both as arbitor and practitioner.

Summary

This chapter has placed YOU, the researcher, front and center as the most important resource you will manage. It calls for attention to vertical and lateral processes to encourage creative thought. To structure the knowledge base for problem solving is itself a creative process. Given the power to reason and explore the great potential of the mind, you can achieve well beyond current understanding. Doing so, however, requires an ability to balance the many logical activities of the conscious mind with the creative quest of the subconscious mind. To manage this potential in the context of collaboration within research and learning environments is, if not your responsibility, at least a viable choice. Your responsibility, however, does extend also to the ethical implications of your process and your product.

3

Theory and Applied Research: An Interactive Relationship

Theory is necessary to applied research and applied research contributes to the body of theory.[1] Without discounting the value of practice and experience, the greater the command of theory possessed by researchers, the broader will be their capabilities and the more efficient they will be in planning and conducting applied research projects. This is true because human reasoning and empirical understanding, structured in theory, envelope and support the entire research process.[2] Without a good command of stress theory an engineer cannot properly design nor efficiently build a safe bridge. Plant breeders and biogeneticists must understand the theory of genetics before they can hope to efficiently develop strains resistant to certain diseases. An agricultural economist

---

[1]For literature concerning the role of theory in research see Brauer and Hunter; Kmenta; Ladd; Leontief; Allin; Bear and Orr; Halter; Krupp; Kuznets; and Nagel (1961, 1963).

[2]Kaplan presents an excellent perspective on inquiry and mentions one of the major contributors to scientific methodology, Immanuel Kant. Kant's contribution to inquiry continues to influence contemporary thought and science, particularly his approach to conceptual work. Kaplan's work draws key elements from Kant as he observes: "Since Kant, we have come to recognize every concept as a rule of judging or acting, a prescription for organizing the materials of experience so as to be able to go on about our business." (p. 46)

"It follows that concept formation and theory formation in science go hand in hand -- another great insight owed to Kant." (p. 52)

cannot determine an optimum farm organization without knowledge of production economics theory. The cultural anthropologist relies on various theories about households, families and other collectives when contributing to medicine, agriculture or poverty alleviation.

The researcher's foundation in theory provides the orientation for defining a problem that is researchable within the discipline or disciplines involved in the research and with the resources available. Theory also provides the basis for the formulation of hypotheses and in the selection of the analytical techniques to be used. And it should be obvious that the interpretation of the results depends heavily on the theoretical orientation of the researcher. These statements are true even in integrated, multidisciplinary research as in agro-forestry and farming systems research.

Although theory permeates the entire research process, in applied research, frequently conducted under sub-optimum conditions, the researcher's practical experience is equally important. Institutional and budgetary constraints, less than ideal field conditions, poorly trained personnel, inadequate background information and other limiting factors have a very significant effect on the research process and therefore must be recognized and dealt with accordingly. Appropriate and properly utilized theory can help attend to some of these constraints depending upon the nature of the problem under study.

Functions of Theory

Maintaining a balance among philosophical, methodological, theoretical and information gathering orientations is necessary for problem oriented research.[3] This balance gives scientific research methodology a creative

---

[3]Kant again makes a major contribution extending to contemporary thought and indirectly to practice. For the dedicated, a study of the Prolegomena is recommended. The *Prolegomena* by Kant (1724-1804) is considered to be a classic in metaphysics and the theory of knowledge. For the student of philosophy it is a companion to the *Critique*. An important contribution to methodology is found in the three conceptual models for understanding a priori principles: Logical Table of Judgements, Transcendental Table of the Concepts of the Understanding, and Pure Physical Table of the Universal Principles of the Science of

flavor, but one based on problem orientation evolving from felt needs. Theory plays an important role in research by providing structure and form to "known" relationships as a basis for developing additivity and efficiency in management of the knowledge quest. Theory provides a sustainable basis that nurtures the progressive and necessary accumulation of knowledge to solve problems and create opportunities. That is, with theoretical grounding we need not begin completely anew with each new problem. Yet often of necessity, conceptual adaptations are drawn from, but complementary with, various theories. These adaptations serve as evolving guides to applied research and may in time contribute to refinements in the theory.

> Theory provides structure and form to "known" relationships.

Classification

The processes of evolution, development and use of theory are normal and necessary within human reason. Theory provides the "language" and "filing system" for conceptualizing relationships (systematic behavior) and classifying information (facts and conditions). Facts, without classification, can not be understood and managed. Experience, transmitted through conceptualizations of need, order and causality, guides us in selecting from alternative classifications of fact. These conceptualizations have their roots in theory; they also provide for further development of dynamic or evolutionary theories.

By depicting the development of a system of relationships, theory permits us to conceptualize the interactions within a system. The theory of supply and demand serves as an example. Into these two categories we group factors affecting prices of products and prices of production factors. Items which affect prices are income, population, factor prices, production coefficients and others. The first two include demand relationships while the last two are supply oriented. Similarly, plant production is influenced by genetic characteristics of the plant, soil and water conditions, and cultural practices, all containing a multitude of factors grouped according to numerous theories that are nurtured by

---

Nature. His desire is to explain how pure mathematics, science, nature and metaphysics are possible and give the bounds of pure reason.

various disciplinary and multidisciplinary configurations. The research process and analysis of research results flow from these theoretical formulations.

Research performance can be influenced greatly by the quality and applicability of the theory employed. The applicability theorem is often responsible for economists being called subjective. For example, facts to be studied or statistics to be computed sometimes can be selected in a biased manner. The type of system assumed causes the researcher to look at one component or relationship instead of another. For example, in a competitive market the existence of stable equilibrium solutions may be assumed and effort is devoted to exploring stable demand and supply functions. But if the competitive market is viewed as "anarchy", it may be assumed that no equilibrium solution exists and "implications of badly behaved supply and demand functions" might be explored.

Specific contributions of theory to understanding a system of relationships can be elaborated:

1. Theory provides clarity and precision to thought systems by encouraging efficient use of the scientific method to enrich and, at times, replace the evolving experience of repeated common sense findings.
2. Theory gives a structure for the prediction and/or measurement of "facts."
3. Theory points to knowledge gaps, limits and deficiencies regarding both facts and relationships.
4. Theory permits concise summarization of what is already "known" about the object of study through both empirical generalization and a system of relationships (the former may become accepted as supporting laws and the latter as the theory).

Evolution of Theory

Development of theory is a natural, nearly involuntary human behavioral trait that permits growth of the individual in succeedingly comprehensive strategies for self management and development. And theory plays a similar role for groups of people with common concerns. Knowledge results from the consensus process that leads from fact to law to theory (Figure 3.1). A fact is an empirically verifiable observation.

If well established, one set of factual relationships can be generalized and, when broadly accepted or tested, becomes law. Theory refers to sets and systems of relationships among laws.[4] Expansion and application of theory results in the identification of various sets of relations among facts. So facts are initiators of theory and serve as a basis for relationships in the establishment of laws, thereby directly influencing the validation of theory. Caution is necessary: as facts change, relationships change; relationships among erroneous "facts" are erroneous and result in weak if not unacceptable theories. Thus stated, relationships are not easily established and require much testing and interpretation. Scrutiny of the process and results of observing facts and relationships (hypotheses, generalizations and laws) is required if theory is to have the conceptual force necessary for assisting with applied research.

> Humans naturally develop and test theory from observed or experienced relationships.

Law, as the intermediary between theory and fact, is an accepted generalization or set of generalizations resulting from successful inquiry through testing of relationships (hypotheses). Laws, once set forth, provide a basis for:

- identification by discriminating among differences in related factual situations,
- presupposition about appropriate data and generalization,
- supposition in the form of assumptions that are not necessarily given by experience,
- hypothesis specification to establish the mode of inquiry,
- linkage among systems as principles to engage reality, and

[4]Hausman introduces the philosophy of science in a cogent statement about theory, laws, observation and unity. In this work Hausman assembles many of the major texts. Of interest here is the direction some of Kant's work took as a basis for complementary or opposing views among classical economists. Logical positivism found efficacy in economic, religious and other orientations in thought extending beyond the classists. Represented among numerous writers are J.S. Mill, J.N. Keynes, Weber, Marx, Veblen, and J.M. Keynes.

- as resources or tools in a procedural context to guide measurement.

Validation of theory, usually not an "all or none situation," often includes only some of the embodied relationships yet results of the validation process reflect on the entire theory. A theory is validated by correspondence with factual situations, by coherence relative to a larger body of theory and/or by pragmatic value in application.

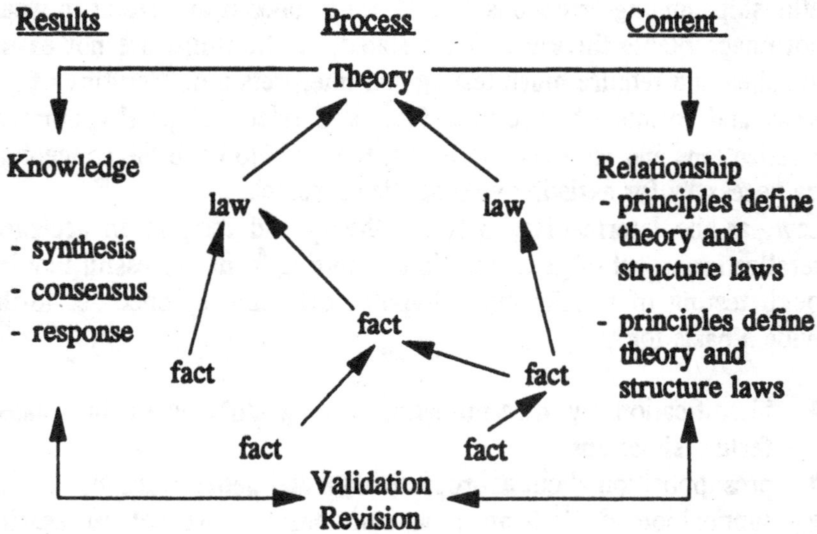

FIGURE 3.1 Conceptual presentation of the relationship between theory and fact

Figure 3.1 provides a conceptualization of theory, law and fact as a basis for a reductive approach to managing knowledge. Theory is most useful when we have developed a "theoretical system" which consists of an organized set of relationships (principles). This organized set is what Kaplan refers to as theory. When we have only one law or generalization then this is our theory.[5] When we have various generalizations to build from we have an organized system of laws for our theory. Principles provide the organization for managing the evolution from observed relationships of facts to laws and laws to theories. The role of principle in the theoretical system takes the form of assumptions, conventional beliefs and other "rules" that condition the environment within which theory evolves. The principles assist in specification of hypotheses which express relationships and, when verified through application and time, serve as the basis for generalization, law and theory. Knowledge formulation interacts with the evolution of theory and involves consensus to synthesis resulting from validation of various responses to the interaction of facts.

An example of a theory and a selected set of some of its components, as conceptualized above, might include the following:

Theory	Neo Classical
Theory/subset	Supply
Principles	Marginal, Average, Fixed, and Total Cost Relationships
Law	Diminishing Returns, Comparative Advantage
Principles	Marginal Rate of Technical Substitution, Production Function Relationships
Generalizations/Propositions	Input/Output Information
Facts	Input Data, Agro-climatic Data, Output Data

[5]Kaplan, as a philosopher, systematically enters the inquiry processes of behavioral science. As a result *The Conduct of Inquiry* serves as a methodological starting point for behavioral science research by giving currency to a social science. Kaplan draws from various philosophical traditions, including Kant, particularly with reference to cognitive means for knowledge retrieval as well as to the processes in science of concept and theory formation.

Theory and Knowledge

Knowledge, being more than information, is based in a system of laws more commonly viewed as theory. It is not a collection of laws based in various unrelated but proven hypotheses and generalizations; instead theory in support of knowledge is a system of laws brought together as relationships that can be altered or expanded through evolutionary hypothesis testing and generalization. The recent emergence of thought surrounding farming systems research is an example of an evolutionary system of "laws" flowing from socio-economic and bio-physical scientists who desire to better understand and "assist" primarily small farm agriculture. Post harvest food handling and distribution is assisted by many theory based systems including physiology, pricing, transportation, storage, processing, inventory, promotion, quality control, consumer preference, and others.

Theory, or laws within "our theory," are used for orientation and as an organized system of relationships that narrows the range of facts to be studied.[6] A dilemma emerges where an "optimal" balance of activity in fact-finding is necessary. "Perfect knowledge" does not result from measuring everything because, even with advanced measurement technologies and computers, we can not comprehend everything. Thus we settle for priority facts and a measurement of only a portion of the relationships among priority facts. Considered in the context of a learning curve as depicted in Figure 3.2, we note that neither stages I

> Theory extends the learning process and the knowledge base.

[6]From Walsh (p. 313) writing about Kant: "...the central problem of the analytic: How can concepts that do not originate in experience find application in experience?...For Kant a scheme is not an image, but a capacity to form images or (perhaps) to construct models. ...There must be some connection between the abstract idea and the experienced world to which that idea is expected to apply; it must be possible to specify the empirical circumstances in which pure concepts of the understanding can find application. Kant thinks that for the categories this requirement is met by the fact that we can find for each of them a "transcendental scheme," which is, he explains a "transcendental determination of time."

nor III may provide the best amount of usable knowledge. In one instance there are too few facts and relationships (stage I) and in the other there may be too many and the costs are high (stage III). The optimum amount of information considering both cost and sufficiency is somewhere in stage II. The magnitude of stage II will vary according to the particular case; it is bounded by an information sufficiency requirement for meaningful interpretation and analysis (stage I is insufficient) and the point where resource and time costs exceed the expected value of the information (stage III is not cost effective).

Theory, can help establish cost effective hypotheses for applied problem oriented research. Resource constraints can be reduced where theory, based in past experience and proven conceptualizations, substitutes for investments that otherwise would repeat previous work. Theory and experience both can help in specifying the amount of stage II knowledge required for assisting with the resolution of applied research problems. Theory, while giving data factual status by guiding classification, reduces the need for extending the demand for information beyond stage II to stage III.

While theory contributes to extending the range of stage II in the learning curve model, it does not function without support from a well

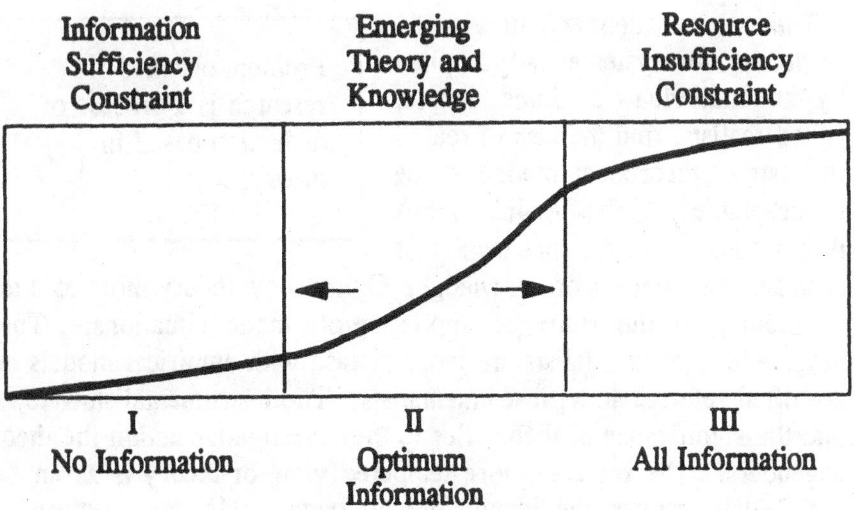

FIGURE 3.2 Learning curve applied to information needs and research resources

established methodology. Part Two of this text is devoted to planning and conducting applied research. Empirical methods are found elsewhere in the literature. As research begins, stages I and III are coterminous; that is, stage II does not exist. The research process, through efficient and effective methodology and research resource use provides a conceptual framework from which theory begins to emerge so that stage II of the learning curve can take form and expand.

Thus, the process by which theory provides meaningful knowledge destined to problem oriented research, and depicted by stage II in Figure 3.2, is one of adding structure (and hopefully cost effectiveness) to the classification of fact and to the conceptualization of factual relationships. Formulation of generalizations (often stated initially as hypotheses) about factual relationships may be efficiently guided by theory. Such generalizations can form the basis for establishing or altering widely accepted sets of generalizations (laws). Theory then, through research applied to real problems, becomes a resource for interpreting, criticizing and unifying established laws that have empirical and experiential content. Herein lie the rudiments of a dynamic knowledge base.

Application of Theory

> Problem oriented research is a process of deduction based in theory.

Theory is neither universally respected nor systematically applied by all individuals. Some people, called realists, find the lack of reality in abstract theoretical models to be unacceptable. The realist views theory as a picture or map that sketches the frontiers of knowledge. Others view theory more as a tool of inquiry in the study of applied problematic situations. These pragmatic-experimentalists are more at ease with empirical models and are often referred to as instrumentalists. The instrumentalists' view is that the significance of theory lies in the investigative action the theory engenders. The realists' more tempered view of theory is as an "art form" that inspires the imagination to better understand reality. In practice this duality seldom prevails and "good" research requires some of both.

Formulation for Application: To and From Hypotheses

Theory formulation and application is initiated through the hypotheses. When facts and relationships are assembled, ordered and conceived in a relationship, they constitute a conceptual system that may become a validated theoretical system. The various facts and relationships in a theoretical system may be logically analyzed and relationships other than those stated in theory, can be deduced. From an instrumentalist perspective, we can deduce specific relationships between specific factors for our specific problems by using theory because it is general. The process of problem oriented research therefore, begins with a problem and usually with a process of deduction in which theory is employed. With the theory we develop hypotheses to be tested in order to find solutions to our problems. Why is it to our advantage to be able to deduce alternative solutions (hypotheses) to our problems from theory? Without theory to determine alternative solutions, we might be compelled to use a trial and error approach and might have to test and analyze every conceivable solution to the problem. This would be taking the realist perspective to the ultimate extreme.

Identifying Theoretical Hypotheses

An hypothesis is a proposition derived through deduction. Propositions can be tested to determine their validity. If they are verified they become part of a future theoretical system either as an empirical generalization or as a law. As an empirical generalization they may refer to a narrowly defined problem. The formulation of hypotheses is explained by stating some difficulties that are encountered during specification as well as the characteristics of usable theoretical hypotheses.

1. Constraints in formulating hypotheses:
    - Lack of theory from which hypotheses may be developed.
    - A poor understanding of the theoretical system or an inability to use the system logically to form "good" hypotheses.
    - Lack of knowledge regarding available research techniques necessary for a proper specification or testing of the hypotheses.
    - Inappropriate use of assumptions in hypothesis formulation relative to unobserved facts.

2. Characteristics of usable theoretical hypotheses:
   - Must be conceptually (theoretically) clear.
   - Must be subject to empirical testing in the sense that they contain no moral judgement.
   - Must be specific: the conditions under which it is expected that an hypothesis will be valid must be spelled out (i.e., Demand for what grade of beef? Most profitable under what conditions? Response to nitrogen in what environment? etc.).
   - Should be related to available techniques (if not, they can not be tested).
   - Should be related to a body of theory.

Integration

> Theory formulation and application are initiated with the hypothesis.

Hypothesis specification is a task that brings theory to bear on applied problem oriented research and vice versa. The causality specified by problem identification establishes the need for research and the causality presupposed by the related research hypothesis helps resolve the problem. Both depend on the body of knowledge expressed within the theory to suggest alternative causal situations from past experience that may lead to successful research and resolution. But simultaneously something may be gained from the test that will respond to the problem and to refinements or further verifications of the theory. This marriage between theory and problem solving is the crux of the natural (or research) process of inquiry followed by humans and the basis for change in knowledge and application.

Summary

Use of theory to advance the quest for knowledge as a basis for dealing with life's problems is a practice that probably began in antiquity. It is so because of our gift of reason, a natural process now assigned various titles, definitions and explanations. Through the process of compiling and forming theoretical systems it is possible to achieve both efficiency and additivity for purposes of confronting ever more difficult problems and challenging opportunities.[7] Careful inclusion of the experience and knowledge in theory will assist us in achieving and sustaining effective and efficient problem oriented research. We now move on to discussing that task.

---

[7]Kaplan may best summarize this chapter "...laws have systemic meaning just as do concepts; just as concepts implicate laws, so do laws implicate theories. From the standpoint of a purely philosophical theory of knowledge, this view may be traced to Kant, with his insistence on the interplay on every act of cognition of the faculties of intuition, understanding, and reason." (p. 99)

"Concepts, judgements, and inferences are as determinative, as Kant elaborated in some detail." (p. 297).

"How we put the question reflects our values on the one hand, and on the other hand helps us determine the answer we get. If, as Kant said, the mind is the lawgiver to nature, it also has a share in facts, for these are not independent of the laws in terms of which we interpret and acknowledge their factuality. Data are the product of a process of interpretations." (p. 39)

PART TWO

Planning Applied Research

4

Effects of Resource Availability on Applied Research

The relationship of research activities to the availability of research resources is an important difference between applied and basic research. In much of what is commonly considered basic research, a proposal is prepared, and if funding is granted the project is initiated. It normally continues so long as financing is available and then whatever can be concluded is presented as the results of the project. Seldom are the results periodically scrutinized to determine their relationship to the proposal because the urgency associated with solving a pressing problem is missing. Nor are results frequently weighed against the use of resources to estimate the productiveness of the project.

> Applied research is accomplished under time, financial, human, information and physical resource constraints.

Applied research is more often (though not entirely) carried out under other circumstances. Because the research is oriented toward the resolution of a specific problem, there is usually a time restraint fixed by the need to make a decision. Such research is also carried out under varying degrees of financial restrictions and often under rather severe shortages of trained human resources and available data processing resources. Other research resources which are seldom abundant under many conditions of applied research are published data, other forms of secondary data or reliable information in general. There may not be an appropriate *theoretical base* that gives structure to known and desired information (Chapter 3 gives a brief statement about the role of a theoretical base). Basic physical facilities such as means of transportation or land area for research can also limit the scope of research activities. Thus, in terms of the learning curve, effective and efficient inquiry (presented in Figure 3.2, stage II) may be very constrained by the availability of applied research resources.

In the process of fitting a project to the resource restrictions, three alternatives are possible; (1) the resources may be expanded to fit the project, (2) the project may be narrowed to fit within the restriction, or (3) both may occur within limits. The first alternative is appropriate if the level of precision or degree of confidence desired by the client prohibits a reduction in the scope of the project. In this case, the client must be prepared to provide additional resources where the limitations are critical whether it be in physical facilities, people, funds, additional time for completion, or some combination of these provisions.

If resources cannot be expanded for a particular project and the deadline for conclusion is firm, researchers have four further possibilities open to them. First, they may study fewer variables, ignoring some relationships which affect the analysis, but those which they hope are less important than the ones included. For example, an agronomist might study nitrogen response for only one plant density and irrigation schedule, rather than for two or more. Second, they may also aggregate variables into groups. One could study the response to a fixed formula fertilizer rather than several combinations of nitrogen, phosphorus and potassium. In this manner, it is possible to include those relationships which may otherwise have been omitted, but the nature of the relationships, because of the aggregation, becomes less clear. A third alternative is to modify or change the nature of the analysis to be carried out. Less complex analyses can usually be conducted more rapidly and with fewer facilities, but the precision of the results is reduced accordingly. Finally, the researcher may choose to make fewer observations either in the form of fewer experiments, treatments or replications, or a reduced sample.

> Decisions are necessary and applied researchers are expected to provide useful information even within limited research resource situations.

It is clear that resource availability has an important effect on the nature of the research product derived and the level of precision which can be achieved or the level of confidence which can be placed in the results. Even in very limited resource situations, however, decisions are necessary and applied researchers are expected to provide useful information. The ultimate decision as to the quantity of resources to be made available for any particular project and the time limit for its completion rests with the client, or with the person responsible for making decisions related to the resolution of the problem toward which the research is oriented. At the same time, clients and administrators depend on the researchers to provide them with accurate measures of resource requirements and the scope and precision they can expect from devoting different amounts of resources to any particular project.

The following sections will describe the principal research resources and include a general discussion of how each resource, when restricted, can affect a project. A single chapter does not adequately cover the range of alternatives open to the researchers and their clients, but we hope that it will provide sufficient stimulus so that the researcher, with imagination, will be flexible in adapting the research efforts to any resource situation.[1]

Information Resources

Information is the foundation upon which research is based; hence, one of the major tasks of the researcher is collection of information for use in the research process. Information in general and data specifically are as critical to the problem identification phase of the project as they are to analysis. Their availability profoundly affects both the quantity and quality of research which can be produced within a given period of time. There are few limits to the quantity of data that can be accumulated given sufficient time and resources; however, requirements are narrowed and brought into focus by careful research planning.

Often, published information provides the basis for problem formulation while either published data or data generated in the research process (or both) may be the source of information for analysis. In some cases, rapid reconnaissance surveys (also called sondeos in farming systems research and extension -- see Chapter 5) are used for problem formulation to augment available secondary data. Published data, or any information not gener-ated or accumulated under the control of the

---

[1]See Arnon; Marshall and Rossman; and Ruttan for discussions concerning allocation of scarce resources to alternative research programs and projects.

researcher but utilized in the particular project are con-sidered to be secondary data. In contrast, any data generated by the researcher and directly associated with the research project are primary data. Another classification of data which has an importance to research is that which differentiates time series data, or observations made at specific intervals over a period of time (for example, experiments conducted over two or more years, or price relationships over time), from cross-section data which are taken at one point in time (for example, experiments conducted at two or more locations or prices from different markets). A third comparison is that between experimental and non-experimental (survey) data. Each of these kinds of data or data sources has different costs associated with time, availability, and analysis, and each has different implications with respect to confidence in conclusions based upon it.

> Data and information requirements are narrowed and brought into focus by careful research planning.

Secondary and Primary Information

When secondary information related to the project is available, its cost to the researcher, both in terms of time and money, is usually less than that required to obtain the same kind of information first hand. However, the usefulness of secondary data as a research resource is not always as great as that of primary information. The researcher must always select the appropriate mix of primary and secondary data as well as the techniques used to combine the two in order to obtain the most useful research product within the limits of the other resources.

For practical purposes, some data are only available from secondary sources. Price series, crop production series and census information are examples. It is simply not appropriate to consider obtaining this kind of information first hand. But it does not follow that available secondary data of this nature are always adequate, representative, or even relevant to the particular project. Researchers must satisfy themselves that any particular series really measures something that is relevant to the project or something which can be made relevant through acceptable modifications or manipulation. If care is not taken to verify secondary information in this manner, the researcher may well draw false conclusions from analysis of the data.

Primary data usually will be more closely related to a particular project than secondary data which are collected for a multitude of purposes or for projects with other objectives. But primary data collection almost always requires more time than is necessary when using secondary data, and may require more of the other resources. Hence, although primary and secondary data are not necessarily substitutes for each other, the researcher should be aware of the availability of secondary information and assess its relevance as an alternative to the collection of primary data.

Time Series and Cross-Section Data[2]

Time series data, as the name implies, are data obtained in a series over a period of time. Examples are price series, regional production and acreage data, and indices of costs of living or wages, most of which must be accumulated over long periods of time to be useful, or crop experiments which are conducted over two or more years to determine the effect of climate. Cross-section data are those taken at a fixed point in time (or over a relatively short period of time) and include observations of several different strata or levels of a population. Examples of cross-section data include those from multi-locational experiments or a one-time sampling of families from different income levels. In many cases, time series data are essential to a project, but occasionally cross-section information can be substituted. When cross-section data can substitute for a time series, the researcher should not despair if a time series is not available. Nor, by the same token, should the researcher blindly use a time series without considering the advantages of obtaining and using cross-section data. Or in some cases it might be desirable to use a combination of the two.

Experimental and Non-Experimental Data

Experimentation is a means of obtaining data with relatively high precision in measurement of the variables. In many instances this precision is associated with a longer time requirement than that needed for obtaining nonexperimental data. In crops, a minimum of one season

---

[2]For a brief discussion of how these two types of data influence demand analysis as one example, see Manderschied.

is required and much longer periods may be needed if, for example, some weather effects are to be determined. A year or more may be common for some animal experiments. When, as discussed in Chapter 1, the Minister requested that the researcher check with someone at the experiment station to help resolve the fertilizer demand question, she was hoping that secondary experimental data might provide specific crop requirement guides to be used in preparation of a demand estimate. She knew time would not permit the design and analysis of crop experiments.

In some cases, however, experimentation can reduce the time and resources required to resolve a particular problem when compared with non-experimental data collection. For example, an experiment (a taste panel for example) to determine potential consumer acceptance of a new product before it is marketed can be less time consuming, require fewer resources and involve less financial risk for an industry than consumer response survey research following the full scale production and marketing of the product.

Yet in other cases, a possible alternative to field experimentation is the collection of non-experimental data through a survey of a number of people who have knowledge or experience with the phenomenon in question. This procedure usually requires less time, but may be less precise and more costly than experimental data collection. An approximation of a fertilizer response experiment for example, can be made by surveying a group of farmers, each of whom uses different quantities of fertilizer. Obviously, results will be *less precise* than experimental results, but at the same time, estimates of variance will be *more realistic* than from a controlled experiment, and the results may *more accurately* reflect farm yield levels, so that farmers will have a better idea of the range of response to expect from the use of fertilizer.

Another non-experimental source of information in the context of applied research is the simulation of results through the use of "best guesstimates" of the most knowledgeable people available to the researcher. For instance, sufficiently detailed input requirements, yields, and resource restrictions can be generated in this manner for use in preparing budgets, and ultimately a linear program or simulation model for the agricultural sector of a region. Such a model, while less valid than could be possible under more optimal conditions, can be used successfully in project planning where time limits prohibit experimentation and current conditions in the area will not provide the detailed information needed from a survey.

Human Resources

As with most resources, the human element must be considered from the points of view of quantity and quality (Chapter 2 views the researcher, a manager and practitioner, as the key resource to successful applied research). Sheer availability is not sufficient for most research undertakings; the training and capabilities of personnel must be considered when planning the project. Except in rare instances, the time factor in applied research prohibits the training of professional personnel though there may be time to train some non-professionals such as interviewers. Field laborers who cannot read or write may be willing to do the physical labor of an experiment, but they cannot always be relied upon to maintain records of the results. Nor can professionals with a minimum of training in statistics be expected to carry out complicated statistical analyses. In the first case, other arrangements will have to be made, and in the second, less sophisticated techniques will have to be employed.

A common misuse of research resources exists where comprehensive data collection processes are employed but the data are not fully utilized because appropriately trained personnel are not available to make proper or correct analyses. In the short run, simpler experiments and surveys which can be analyzed readily by available personnel are more appropriate. By matching data collection techniques and processes to the other available resources, research and development activities can be effective and improve over time. Money saved by not conducting overly elegant data collection programs than are necessary may be used to train personnel to conduct and analyze more complicated and sophisticated experiments and surveys in the future.

Physical Resources

Non-technical physical resources include transportation, land, office space, machinery, word processing equipment, and other items of a similar nature. Like all other research resources, their availability must be considered when planning the project.

Technical physical resources include scientific instruments, calculators, personal computers, etc. Certain kinds of instruments may be indispensable for particular aspects of a project while some hypotheses may need to be eliminated if the correct instrument is not available or is too costly compared to expected results to justify its use. On the other hand, a desk top computer, while not indispensable, may substitute for other resources such as money or time. The use of a computer,when available, sometimes can shorten the time required to achieve useful results. If a researcher must wait long periods of time, however, for data entry and verification and for programs to be de-bugged, he/she may be better off to undertake appropriate analyses on a hand-held calculator. When computers and calculators are unavailable a diligent researcher might still provide rough but meaningful recommendations based upon experience and imaginative use of the most basic of physical resources -- a pencil and paper.

Financial Resources

This resource to a certain extent can be substituted for all the other resources. At the same time it may not be appropriate, feasible, nor even possible to make this substitution because of other considerations. For example, in some instances money will not buy time. Researchers, with some exceptions, must wait for another cropping season, another Christmas season or the next eclipse of the sun for certain observations.

Funds are required for almost all research projects and their availability is an important consideration in research planning. A timely release of funds provides for adequate anticipation of research resource applications to appropriately meet research needs. Good planning and budgeting are necessary to aid in the efficiency and flow of financial resources.

Time Constraints

Time is usually not thought of as a resource in the same terms as physical facilities, information or human resources, but its effect on the planning and execution of applied research is similar. Considered as a resource, time can interact with other resources in that substitution of one for another can be made. If a decision on a particular problem is critical and time is limited, a greater number of other resources will be required to achieve a given level of confidence than would be necessary if more time could be taken. Hence, the use of more time is an effective substitute for quantities of other resources. But also, consuming more time on one project reduces the amount of that limited resource which is available to help resolve other problems.

> The urgency of problem resolution in applied research makes time an important resource.

In some situations, time can be overwhelmingly limited. This is usually the case when one is involved in so-called "brush fire" research of the type frequently faced by planning groups within the various ministries and agencies of government. In such instances the researchers must always maximize the efficiency with which they use the time resource. Only data readily available can be used and lengthy methods of analysis cannot be considered. Often "best guesstimates" are the only means available to the researcher under these conditions.

Approaching the other extreme are theses and dissertations at any of the levels at which they are written in various education systems of the world. Often they are not oriented toward any particular problem so the time factor is not relevant, but among those that are problem oriented, too often time is relegated to a secondary role in resource utilization. As a result, many theses and dissertations are not written in time to be of any great use, their value reduced by failure to account for the time factor. Those who excuse this fault by emphasizing that theses and dissertations are only meant to be training tools often deprive the student of an opportunity to perform and benefit from meaningful research.

Research Resource Planning

All of the necessary research resources ultimately must be accounted for in a plan. This plan accounts for a methodology, for the entire process from problem identification through presentation of results, a budget of financial and other resources that has considered possible substitutes, and a calendar for allocating time to the individual activities and the total research endeavor. The methodology, budget and calendar are interdependent parts of the whole research resource plan within a research proposal.

Various budget formats can be followed, often depending upon the desires of the funding source or client. To the extent that the budget can be developed in response to the research objectives, a process for scaling funds to meet priorities will be more easily implemented. Balance in allocation of funds among different aspects of the methodology as well as to the specific research objectives, assures the ultimate delivery of a product to the client. Careful allocation will help the client determine how much financial support is necessary to achieve a specific information need. How far does the client need to move up that learning curve? What degree of confidence is necessary in the test of hypotheses? Each increment is usually associated with increased costs. Expected benefits will help determine how extensive the research resource investments should be.

> The methodology, budget and calendar are interdependent parts of the whole research proposal and resource plan.

Time planning may involve any number of tools from a simple calendar to very detailed guides. Time can be charted with PERT, a CPM[3], various flow charts and similar mechanisms depending upon the complexity and magnitude of the research project or program. We do not discuss these tools in detail as they are covered in the companion text by McDermott and Andrew. We do emphasize that the effective use of time in research involves a carefully planned process that is served by a well developed methodology. The time plan and the methodology are necessary conditions and go a long way toward cost effective research.

Summary

Research undertaken specifically for the purpose of obtaining information to help resolve a particular problem is applied research. The purpose of this chapter has been to examine the relationship between the scope of an applied research project and the quantity and quality of research resources which can be devoted to the solution of the problem at hand. The urgency of problem resolution in applied research makes time an important resource or constraint which interacts with the other theory, financial, human, physical, and information resources.

---

[3]PERT (Project Evaluation and Review Technique) and CPM (Critical Path Management) are network analysis tools developed by engineers for computer aided time management.

Researchers must be aware of the effect that certain resource limitations can have on their research. This cognizance will improve their research efforts by increasing the probability that the proposed projects will produce useful results. Projects designed in the absence of this consideration can and frequently do run into difficulties -- often the productive potential of the resources utilized is not attained. The result is that less effective information is made available for decision making and problem resolution. Careful consideration of resource availability can help prevent situations where the deadline arrives and the researcher, still engrossed in gathering data or making analyses, has little of value to report to his/her client.

5

Sources of Information for Initial Problem Specification

The lack of reliable information often leads to insufficient problem identification. Without adequate information untested relationships may be assumed or conventional wisdom may be used, both imposing limitations on reliability of the research. Improved methods for collection of information can help with this situation. Recent developments in research methods, for example, by the social sciences, provide numerous tools for open-ended and holistic information gathering. Information obtained through use of these research methods provides the basis for improved problem specification and the basis for greater accuracy in the problem solving research to follow. This chapter considers open-ended approaches from the perspectives of both rapid and prolonged information accumulation. We consider also the use of secondary data and literature reviews to further improve problem identification and enhance the analytical phase of research.

> Open-ended approaches can assist with problem identification for research planning or provide early infor-mation when the time constraint is more immediate.

Open-Ended Information Gathering Approaches

The term "open-ended approach" refers to information gathering techniques that allow for refinements in the methodological specifications of a proposal (researchable problem identification, specification of hypotheses, objectives and possibly analytical procedures), the information gathering techniques and the understanding of research results.[1] Open-ended work often incorporates participation by the informants and the clients more fully in the overall research process than do more conventional information gathering and analysis approaches. In appropriate circumstances both the process and product of the research are enhanced by dialogue. Because the nature of the information gathered may change over time, traditional statistical analyses are not always appropriate. However, because problem identification is enhanced, subsequent research that employs more traditional techniques may then benefit by improved accuracy. Viability and efficiency of the overall research effort should be enhanced by use of a balanced methodology where secondary information and primary information, drawn from conventional data gathering methods, are supported by information obtained from open-ended approaches.

> Open-ended approaches to information gathering capture qualitative responses through participation and dialogue with informants in case studies, focus groups, multidisciplinary teams and cooperative inquiry teams.

Rapid Information Gathering

Various techniques can provide for rapid information accumulation. We discuss focus groups and multidisciplinary team interviews because they are well suited to conducting and planning applied agricultural marketing, production and resource use research. Each is mobilized by management of groups oriented toward a specific problematic situation for which further information can improve problem identification and the intensity of applied research to follow. Besides well managed and frequent reporting practices, the dynamic potential of groups oriented toward rapid information gathering is captured through brainstorming, brainwriting, tree diagramming, flow charting (see McDermott and Andrew) and other additive processes.

---

[1]Gareth Morgan provides an opportunity to peruse several methodologies that might be termed open-ended.

Focus Groups

Focus groups provide qualitative information through focused discussion often by a relatively homogeneous group of people relative to social interaction about a given topic or condition.[2] A purpose is to determine how the selected group responds as a distinct group. Heterogeniety can be gained by subsequently talking with different focus groups each relatively homogeneous but different from the others. The problem situation to which focus is applied must be characterized in such a way that qualitative information is necessary and appropriate. Focus group interaction can help in assessing needs, in identifying and structuring problems and opportunities, testing new interventions, improving existing programs, generating information for more expansive research endeavors (possibly questionnaire development) and characterizing a host of client and customer desires.

Use of focus groups is advantageous when attempting to understand real-life conditions in a social environment. This informant approach provides flexibility, high validity and credibility with target groups, and rapid results at relatively low cost. The approach however, is distinguished from a respondent approach where greater control over individual responses is needed and where greater statistical accuracy is desired for quantitative analyses. Special skills are necessary for management of logistical problems associated with organizing, convening and leading focus groups. Focus group information can be difficult to analyze when major differences appear between different groups working on the same problematic situation. Nevertheless, qualitative insight obtained in an orderly or structured way through focus groups can improve research program planning, design and evaluation. Some researchable problems can be conveniently resolved without extensive research by use of the information obtained in focus groups. This permits the researcher to move on to more specific problems that can be dealt with given the new understanding of causality.

The process of focus group management starts with a group of people who possess certain homogeneous characteristics called for by the nature of the specific task to be undertaken. These people provide information of a qualitative nature in a focused, semi-managed discussion. Typically

---

[2]For a complete coverage of research based in the focus group approach see Krueger.

the group is composed of six to ten people with knowledge and/or experience in the area of interest. The researcher/moderator, accompanied by someone to record the findings, stimulates open group discussion and encourages each participant to influence others as a means to an in depth information search. Thus in an open environment the process taps experience, insight, knowledge and creativity through group work. This information from one group may become input for another group as a means for validation or for continuing an evolving subjective analysis. The product may be a research plan that will initiate a more quantitative effort with confidence that problem specification and research design are appropriate to a client group. Another product may also be deliverable information for ready access by decision makers in public or private organizations.

> Focus groups provide qualitative information through focused discussion by a relatively homogeneous group of people.

As in survey research from a questionnaire, question development for focus group research requires careful planning and implementation. A list of guides to question development offered by Krueger (p. 60-67) includes:

1. Using open-ended questions,
2. Not using dichotomous (yes-no) questions,
3. Rarely using "why" because it delivers quick and much more limited responses than the less directive "what" or "how",
4. Preparing questions carefully and systemically -- brainstorming sessions to develop the question set,
5. Preparing the respondent or participant with consistent background information to minimize tacit assumption,
6. Presenting the questions within a focused context by focusing the topic into the past,
7. Arranging questions in a focused sequence,
8. Beneficially using serendipitous questions usually saving their discussion until last, and
9. Pilot testing the focus group interviews -- accounting for informant characteristics, test the questions with a group (hopefully experienced with focus groups) before performing the first focus group interview.

Multidisciplinary Team Interviews

The primary principle behind multidisciplinary team information collection is to deal with the holistic nature of the informant's operational environment. The farmer, for example, does not think exclusively as an agronomist, an animal scientist, an agricultural economist, a social scientist, or an even more specialized marketing specialist, resource economist or soil physicist. The farmer manages farm, household and nonfarm involvements within the context of a whole livelihood system. A purpose of interviewing a farmer is to gain knowledge about farmer-identified constraints and problems without falling into a "provincial syndrome" sometimes common to academic disciplines.

Multidisciplinary team interviewing can be applied in various situations. A multidisciplinary team interview can help refine problem definition and complements other primary or secondary information retrieval efforts. An explanation of the characteristics common to a multidisciplinary Sondeo team interview approach will help relate this approach to other information retrieval alternatives [Hildebrand, 1981]. Characteristics of a multidisciplinary team interview approach applied to farm families follow:

1. Multidisciplinary interview teams contain members who are heterogenous with respect to discipline so that the holistic nature of the varied enterprise, farm and household situations can be understood. The team provides for holistic information collection and analysis from a relatively homogeneous informant group. That is, informants for a team often include a broad range of farmers but not farmers, retailers, and processors. It is the topic or research problem that dictates the particular characteristic of homogeneity. Note that in the focus group, the group is comprised of the informants or they may be clients. Here, however, the group is the research team.
2. The purpose of the team is (a) to understand the farming system without letting a single professional discipline or subdiscipline overpower the system's interactions that define the whole, and (b) to encompass the constraints and problems for joint (team and farmer) problem resolution.
3. The multidisciplinary team approach to information gathering is both qualitative and quantitative as it attempts first to fully understand the informants' situations (farm and family, resource

base, cultural setting, etc.) and then to measure appropriate indicators of production and management.

Advantages of the multidisciplinary team approach lie in the qualitative and holistic strength provided to aid in identification of informant needs. This then leads (within the farming systems methodology) to further research (multi-disciplinary diagnosis and design of tentative solutions as discussed in Chapter 8) and collaborative evaluation of alternatives within the farmers' environments. Acceptance of improved technologies, practices and policies by the farmers covered by the research is enhanced. A challenge to the research resource base is where the participating target population is so heterogeneous that commonality across the population, particularly in reference to the constraints or opportunities under study, narrowly limits the overall effectiveness of any specific intervention. The approach does attempt to identify relatively homogeneous recommendation domains following from the team interviews and the analysis. The multidisciplinary team approach tends to perform very well in delivering problem solving technologies, practices and decision guides regardless of the size of the client population.

For success of a multi-disciplinary team effort, management and administration must account for the different research resource requirements for information gathering. Compared with either experimental research located at an experiment station or large field interviews from sample surveys, the multidisciplinary team works within different time and spatial constraints. The following lists of survey methods and prerequisites for multidisciplinary team interviews, while not necessarily inclusive, do suggest a need for adaptive management and administration.

Survey methods for multidisciplinary team interviews are numerous and evolving. Several considerations are included that apply to farm level research:

1. Usually a core survey team consists of social scientists paired with biophysical scientists. Disciplines may be represented according to dictates of the problematic situation and the research resources available.
2. The team may include about five pairs.
3. Pairings of interview partners are changed daily to reduce interviewer bias and to increase cross-disciplinary interchange.

4. The team of ten or so people meets usually in the evenings to make tentative interpretations and modify questions as needed.
5. Each team member contributes his or her own disciplinary specialty but subordinates to the common objective of fully understanding the informant.
6. The team may also assist in collecting time series information such as farm records from the informants.
7. The team, in collaboration with the farmer-informant, may become involved in interdiscipinary problem solving research, such as on-farm experimentation, to develop useful technology.

Prerequisites for successful multidisciplinary team interview work follow:

> The multidisciplinary research team obtains information from informants as a basis for holistic responses to client needs.

1. Team members should be well trained in their own fields but need a working understanding and desire to contribute to other participating fields as an integrated whole.
2. Multidisciplinary team members should be encouraged to feel that it is not necessary to defend individual disciplines but to defend the goal of a successful inter-disciplinary effort.
3. The product of a multi-disciplinary team should be viewed as a joint effort with proper recognition for the *team*.
4. The team should be team-product oriented, consider a wide range of variables and constraints (not leaving the broad issues to one person), and spend both field and desk time as a team to gain optimal integration of systems observations.
5. Emphasis on team work and not management and membership, as in a committee, is essential -- integrated, collaborative and joint participation is desired. Of course good overall coordination and management, while participatory, is essential.

Extended Information Gathering

Research is a continuous process. If continuity can be maintained, as a dictate of client need and not methodological self interest by a researcher, the search and research process will be more effective. Several techniques provide for this continuity. One, cooperative inquiry, is both a management technique for bringing diverse experiences and desires together and a powerful tool for bringing problem specification to a higher order of holistic identification for further specialized research. Cooperative inquiry is both a monitoring and planning technique that is best achieved over an extended period of time but it also results in a continuing flow of researchable problems. Another technique, selected to provide contrast, is that of the case study. Here depth is provided by characterizing, analyzing and manipulating a case situation for extended learning purposes.

Cooperative Inquiry

Cooperative inquiry is another participatory research approach, sometimes termed action science, that involves collaborative, creative and directed learning by researchers and people related to the situation under study [Reason, pp. 18-39]. The process is one of extended dialogue and holistic understanding from extensive non antagonistic, yet critical subjectivity that culminates in group learning. This pluralistic endeavor draws on different social, employment and experience orientations by doing research *with* people (participants) rather than *on* people (subjects). This distinction is a matter of degree relative to other approaches discussed in this chapter but it should be noted that cooperative inquiry expands the range of participatory research.[3]

The cooperative inquiry team may consist of three or four people or may expand to a core of twenty. Size, of course, influences management and time requirements owing to the geometric increase in interactions as group size increases. Yet the chance for successful discovery may be augmented many times by the wider array of available experience in a relatively large and well managed group. Time may be of minor consequence in situations where a "think tank" continues to stimulate individual efforts into progressively greater work beyond the confines of

---

[3]See W.F. Whyte on participatory action research methods.

the group. The group itself, however, may very well contribute directly and expediently to resolution of various problems. Contrary to the focus group, where informants or clients are relatively homogenous, and multidisciplinary teams where academic disciplinary participation by researchers is heterogeneous, cooperative inquiry relies on a relatively heterogeneous participant base that joins researchers with clients, citizens, civic leaders, business managers, social leaders and others all with a common interest or concern in the problematic situation. This united cooperative inquiry team makes little distinction between the client and researcher. Often advisory boards and committees have some cooperative inquiry characteristics.

The process of cooperative inquiry is one of cycling and iterating to ever more informative results. Participants may meet regularly for several years before the process is exhausted and the results may take many useful forms over time. Of course, the effort may be more intense and time collapsed when a pressing concern necessitates more immediate results. In either case the cycle of cooperative inquiry runs as follows:

Stage 1 -- Discuss and agree on *WHAT* ideas, theories, models, conceptual forms, and research actions are to be followed; and *HOW* to observe, record, and measure experience for reflection.

Stage 2 -- *ACT* by bringing practical knowledge into the process and observe group exchange and outcomes (physical, psychological, interpersonal, or social); and observe behavior individually and as group for purposes of maintaining group creativity and interaction.

Stage 3 -- Encourage *ENGAGEMENT* with the material in an experiential process to discern interactive results by isolating preconceptions and prejudicial influences that appeared in stage one.

Stage 4 -- *REFLECT* on the experience through both cognitive and intuitive forms of knowing; determine what makes sense (sort and discard); specify *WHAT* will reinitiate the cycle with stage 1; and design subsequent cycles and reinitiate the process (if needed).

Participants in a cooperative inquiry team assume several roles within a cycle and over the duration of the research. They are informants and leaders and must commit to full participation if maximum problem solving creativity is to result. Skills required include genuine collaborative ability with peers, a willingness to help manage one's own quest and anxiety as well as that of peers, attentiveness to the resulting experience that emerges from the rigor of conscious challenge and negotiation, and an ability to formulate and validate results. Specific facilitative roles that may be assumed by the initiator as well as all participants, but often by those with unique skill for the task, include:

Group facilitation -- permits, encourages and nurtures expression of different opinions from varied role and experience orientations;
Distress facilitation -- strives for new understanding and new processes (could be lateral thought as discussed in Chapter 5) by challenging participants to take new risks and expose new ideas which may stress the entire group into expanded insights. (Thus, collaboration itself may prove very difficult early in the effort); and
Inquiry facilitation -- provides for subgroup monitoring of the inquiry process relative to the plan, accompanied by short statements from the initiators to kick-off or redirect the inquiry discussion.

Preparation for cooperative inquiry requires that the initiator be certain that the effort is generally understood by all participants. Understanding about time and resource requirements called for by the group, along with demands on and unique needs of each individual are necessary so that "contracting" can result in consistent participation. An overall plan should be negotiated and agreed to with specific times set for renegotiation.

Validation is important particularly to stage 3 (engagement) in the cycling process of cooperative inquiry. From Peter Reason (pp. 123-125) a checklist is developed that parallels some of the lateral thinking concepts presented within "YOU management" in Chapter 2. It is not intended that the list be sufficient nor necessary for successful work. Rather, it should stimulate individual and group behavior to higher levels of achievement, or to resolving stalemate situations. The checklist of validation procedures includes:

1. Research cycling through reflection and feedback,
2. Balance of divergence in experiential content with convergence toward a whole system,
3. Balance between reflection and experience,
4. Balance between content and process as group work proceeds,
5. Procedures to counter a tendency toward vested interests and collective collusion by a group's desire to save something that should be discarded (or discard something that should be saved),

6. Use of chaos as a learning tool before closing too quickly on order,
7. Management of intrusions resulting from anxiety caused by group interaction,
8. Ability to sustain and nurture authentic collaboration,
9. Skill in anticipating and validating open and closed boundary processes, non-member feedback and experience as inputs (non-member participation must be acceptable to the group and carefully managed),
10. Commitment to sharing viewpoints and in syntheses of alternatives to enable coherent action, and
11. Processes for replicating content of results in an imaginative sense as a test before release (going public).

> Cooperative inquiry is collaborative, creative and directed learning (inquiry) by researchers and people involved in the situation under study. It expands the range of participatory research.

An excellent example of cooperative inquiry is provided by Yee in discussing the 1989 Alar-in-apples incident which resulted in a major food safety credibility gap between farmers and consumers in California. The Ventura County Food Safety Study Group was established consisting of "several lemon, orange, avocado, and vegetable farmers (conventional to almost organic), the president of Mothers and Others for Safe Food, a retail store produce executive, representatives from farm labor, California Rural Legal Assistance, the Farm Bureau, people from the Environmental Coalition, the Sierra Club, the Green's Party, the League of Women Voters, and the manager of a chemical testing lab.

"Our purposes were to reduce conflict, build trust, open communication, exchange information, broaden perspectives, and to study ways to improve the credibility and policies of the food system in Ventura County. In essence, this process opened up the food system and allowed for feedback, input, and ideas from the total community," [Yee, p. 3]. The effort, extending more than one year, accomplished these purposes and improved the linkage between farmers and consumers yet the task was not easy. Yee (p. 4) notes that, "It's apparent from our experience, coalition -- building for problem-solving and consensus is an evolving art on science."

Some cooperative inquiry ventures may be less problem specific in their initial stages than the Ventura County effort. The team itself may play a heavy role in problem identification within a very broad problematic situation.

Case Studies

Case studies may be used as teaching devices, forms of record keeping and as research tools. Each requires somewhat different formulations but may share some common characteristics. As a research tool the case study provides a holistic point for observing the structure and flow of real-life events. Case studies, for example, can focus on managerial processes, village and community change, individual life cycles, farming systems, socio-cultural processes, research processes, institutional evolution and international relations. Yin (p. 23) defines a case study as "an empirical inquiry that:

- investigates a contemporary phenomenon within its real-life context when
- the boundaries between phenomenon and context are not clearly evident, and in which
- multiple sources of evidence are used."

While the experiment deliberately separates a phenomenon from its context and the survey has limited power to measure continuous phenomena and contextual interaction, the well designed case study observes holistic interaction in a contemporary situation. The case study, given the characteristic of holistic interaction, is used sparingly in a large population because complexity and cost prohibit extensive use of this intensive method. Even when multiple-case studies can be designed, their reach remains constrained by limited research resources. It is also difficult to determine if the cases are representative.

> As a research tool, the case study provides a detailed and holistic process for observing the structure and flow of real-life events.

Design of a case study must answer four tests described in full by Yin (pp. 40-45). *Construct validity* is concerned with establishing the appropriate operational measures for the variables to be studied. If one desires to study changes in farming practices, several factors may be relevant and each must be placed in a measurable context. *Internal validity* is concerned with causal relationships where identifiable conditions lead to known results. Hypothetical relationships are not appropriate in this explanatory test because the time-series nature of unvalidated causality could lead quickly to incorrect research interpretations. Validated causality situations become very important as guides to problem specification for most forms of research. *External validity*, to the extent possible, specifies the domain to which the case study findings can be generalized. It, however, is often difficult to determine external validity if the general population is large. At this point replications become an important but added cost of validation. Finally, *reliability testing* determines if the numerous measures can be replicated and thereby indicate that both error and bias have been minimized to the extent possible.

Yin, in bringing together the work of many social scientists provides a check list of considerations for implementing case studies. He starts with required skills for the researcher (pp. 62-63).

- A person should be able to ask good questions -- and to interpret the answers.
- A person should be a good "listener" and not be trapped by his or her own ideologies or preconceptions.
- A person should be adaptive and flexible, so that newly encountered situations can be seen as opportunities, not threats.
- A person must have a firm grasp of the issues being studied, whether this is a theoretical or policy orientation, even if in an explanatory mode. Such a grasp reduces the relevant events and information to be sought to manageable proportions.
- A person should be unbiased by preconceived notions, including those derived from theory. Thus, a person should be sensitive and responsive to contradictory evidence."

Yin (p. 70) also provides a list of required components for a protocol to achieve reliability of case study research. Yin (pp. 85-95) discusses the sources of evidence or information bases for case study research including documentation, archival records, interviews, direct observation, participant observation, and physical and cultural artifacts. Finally, Yin

(p. 106) provides analytical strategies that suggest some familiar techniques.

The case study approach is well suited for unique situations particularly as both a research and pedagogical tool. Adaptations of the approach can be made for incorporation with other information gathering techniques as a worthwhile complementary information source. A particularly appropriate set of eleven farm level cases from seven U.S. states is provided by the National Research Council pertaining to alternative agriculture [NRC]. Feldstein and Poats address gender analysis in agriculture with a case study approach involving seven country cases drawn from Africa, Latin America and Asia. On-farm research cases are provided by Tripp, again taken from nine developing countries. An extended African village case with ethnographic detail is found in Skjonsberg.

Information Retrieval from Secondary Sources

Often researchers have no control over secondary data measurement but, if they are going to use a particular set, they have the responsibility for checking very carefully to determine exactly how the data were derived, the nature of the aggregations made and if the data will meet the needs of the research [McKee]. Secondary sources can be classified into two major types: (1) regular and long-run series that measure phenomena such as prices, income, production, rainfall and temperature; and (2) less regular and, in general, short-run measures of phenomena more often associated with another person's research such as crop response to improved inputs, consumer attitudes, and management characteristics.

Time series data often can be difficult to use and verify because they are secondary and not necessarily designed for a specific research project. Because time series frequently cover extended time periods, collection is performed by an agency which is responsible for maintaining continuity and consistency in the collection process. No particular person is responsible for collecting these data over the years, but certain procedures are established and revised occasionally to accurately and adequately measure the particular phenomena under study. Nevertheless, definitions should be checked carefully so that the researchers can decide what the data in question are supposed to measure, and since they have no control over the collection of these data, they must satisfy themselves that the data in fact do measure the phenomena of interest.

Use of non-experimental secondary data requires researcher responsibility in verifying the quality and applicability of the data to the problem under study.

Comparability may be the most common difficulty in using times series data. This refers to (1) comparability of different time series which purport to measure the same phenomenon and (2) comparability of different periods within the same series. For example, due to plant breeding and food processing programs, the maize produced and marketed fifty years ago and the maize produced and marketed today are virtually two different products. A price series for maize covering this period of time is not comparable during the entire period. Coupled with product changes in a price series is the equally difficult problem of non-comparability of currency values caused by price inflation or deflation in the national economy.

A serious comparability problem, and one that is difficult to detect within a time series, is a change in definition. For example, a change in quality standards will affect a price series, but may not necessarily be specified. A change in the point of measurement such as the location of price determination from the farmer-assembler level to the wholesaler-retailer level will influence the value of the economic variable in question. In census data, one may find a change in the definition of a farm. Similarly, changes in the population of a city may be due in part to changes in the boundary lines.

Non-comparability between series attempting to measure the same phenomenon is common particularly in developing countries because time series measurement systems often have not been standardized under the authority of one agency. Unfortunately, in many cases these entities neither clearly define the phenomena to be measured nor the precise points and times of measurement. And when the points and times are once defined, later changes may not be indicated.

As an example, consider the separate series of potato production data compiled by the Caja Agraria (a public agricultural credit agency) in Colombia and IDEMA (a public agricultural marketing agency) [Andrew, 1969, p. 230]. These series during the 1960s were vastly different, with the production data of IDEMA averaging about 40 percent lower than the data of the Caja Agraria. This difference was probably due to differences in methods of measurement of the series. The Caja Agraria primarily developed its series from producer loan data, while that of IDEMA was based on measures taken in rural markets. Thus, Caja Agraria more closely estimated total production and IDEMA measured the production entering commercial channels. If the researcher is interested in the volume of commercial sales of potatoes, the IDEMA data probably were more accurate. On the other hand, if the researcher wished to estimate per capita potato consumption, including both rural and urban consumers, the IDEMA series would underestimate this variable by about 40 percent less losses and seed requirements.

Where a time series is needed but unavailable or possibly inappropriate for a specific problem, cross-section data sometimes can be substituted. Cross-section data (experimental and non-experimental) represent phenomena measured at only one point in time or a limited number of times. One means of using cross-section data to simulate longevity is in questionnaire research where respondents recall information concerning an event at present, last year, five years ago, and so on. Care must be observed in interpreting these measures because of what has been termed a telescoping bias. That is, the tendency to completely overemphasize or underemphasize certain phenomena by projecting present conditions to previous points in time while forgetting others that may have influenced the observations.

Summary

Varied sources and means for gathering information assist with planning and implementing an applied research program. We do not exhaust the rich range of possibilities. Discussions of rapid information gathering approaches provide a basis for knowing that a great amount of worthwhile information for both problem identification and some of the analyses to follow may come from more qualitative interactions. Focus groups and multidisciplinary team interviews serve unique but, in some instances, complementary purposes. Extended information gathering approaches such as cooperative inquiry teams and case studies provide for on-going observation as complements to other approaches. Often secondary information serves the valuable role of conserving research resources by augmenting knowledge for problem identification and providing data for inclusion in the analytical aspects of research. Careful assessment of the nature, application and appropriate use of secondary information is necessary for every research effort.

6

Orientation and Focus of Projects: Researchable Problems, Hypotheses and Objectives[1]

Proper management of applied research requires clear definition of the goals to be achieved through the project. If researchers do not know for what they are striving they cannot hope to effectively accomplish their tasks.[2] Orientation and focus of an applied research project include the specification of the problem in terms which make it amenable to research, the formulation of hypotheses which are subject to being tested, and the delineation of the specific objectives which the project should accomplish. The interaction of these three phases and their clear exposition serve as a plan or guide for determining the procedures to be followed by the researcher in conducting the research. Adequate orientation and focus of the project will help assure that it can be completed satisfactorily within the resource limitations facing the researcher and will also serve to explain the nature of the undertaking to the client or the administrators for whom the research is being undertaken.

---

[1]An earlier version of this chapter including Figure 6.1 appeared in *Planning and Conducting Applied Agricultural Research* by Andrew and Hildebrand and copyrighted by MSS Information Corporation (1976) and Westview Press (1982).

[2]See references to the mission of research such as West and Salter. Other references are discussed in McDermott and Andrew.

A Conceptual Model

We have found it useful to consider the process of planning an applied research project as being equivalent to a funnel with a series of filters (Figure 6.1). Such a funnel is used to reduce a large volume of liquid to manageable proportions. The orientation and focus of an applied research project serve the same purpose -- they reduce a large volume of information to manageable proportions. Extraneous information and ideas are eliminated as foreign matter might be filtered in the funnel. Each part of the project statement -- problem, hypotheses, objectives -- serves to narrow down the proposal, to bring it into sharper focus, and to filter out surplus or unrelated information, making the orientation more precise. One can consider the size of the lower opening of the funnel as being determined by available research resources (the bottleneck). Hence, the proposal as it finally emerges must fit within these resource restrictions.

The top of the funnel will be the general subject matter orientation of the researcher. The research will be within the area of interest of the individual researcher and usually related to the researcher's individual talents. The general problematic situation will fall within these limits, but any problematic situation suggested by a client may contain several researchable problems. Hence, the selection of a researchable problem based upon the client's needs is equivalent to sharpening the focus on a particular aspect of the more general problematic situation. Hypothesis formulation narrows the problem to tentative relationships which will be tested in the research process. Finally, the objectives specify the limits within which the project will be conducted and describe the useful product which will result.

Obviously, information, ideas and relationships do not flow through the funnel like water but are filtered time and again. The process of moving from the top to the bottom of the research project funnel requires push and pull, pare and adjust, specify and redefine until the proper proportions are achieved. The problem, the hypotheses and the objectives may each have to be changed and adjusted many times before a satisfactory product results.

> A well-conceived plan will increase the probability of accomplishing research to help resolve the problem.

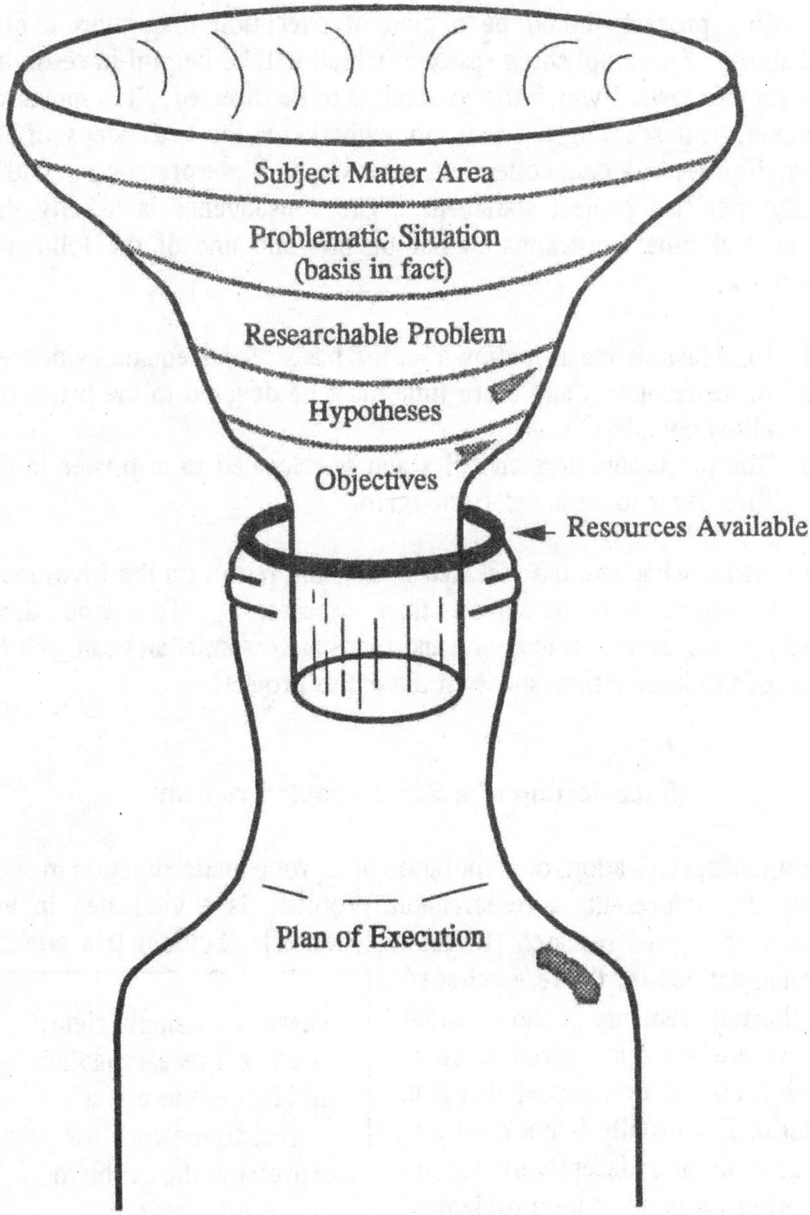

FIGURE 6.1 Fitting the research project to the resources

At this point, a word of caution is in order. The result of this funnelling process should be a plan of execution that gives a high probability of accomplishing research which will be helpful in resolving the problem toward which the research is to be directed. It is much too common, and seemingly easier, to embark on the next steps of the research process -- data collection, analysis, and interpretation -- with a poorly specified project statement. The consequence is usually that budget and time constraints cannot be met and one of the following results:

1. Conclusions must be drawn on the basis of inadequate evidence,
2. More resources and more time must be devoted to the project to allow completion, or
3. The project withers and dies and is relegated to a drawer in the file, never to be heard from again.

No matter what fate the research meets, the return on the investment in the project will be lower than necessary. To avoid these consequences, careful orientation and focus are essential and can well be the most productive time spent on the entire project.

Specification of a Researchable Problem

Problem specification, or elaboration of a problematic situation in such a way that it presents a researchable problem, is a vital step in the process of applied research [Brandow; Weaver]. Seldom is a client's problem defined for the researcher so that the requirements of the research process are obvious. Even in cases where it may at first appear that it is so defined, it usually is not the case. A common and deceptively simple appearing example is the problem of determining crop production costs. An economist cannot uncritically accept a "cost of production" project without understanding which specific cost components are of interest to the client. Researchers, because of their training, will usually have a greater appreciation for the technical aspects of the problem than will their clients and should

> There is no single right, or correct researchable problem -- there is a correct framework for expressing the problem for useful research.

therefore consider it part of their responsibility to identify symptoms and diagnose the problem in consultation with the clients who will have a greater appreciation of their needs and constraints. Researchers should perceive the identification of the problem as a major task, but they cannot assume they automatically understand the problem better than the clients. Researchers must work with the clients until they have jointly defined an acceptable and researchable problem.[3]

Problem specification is not a simple process. Hildreth and Castle summarized a discussion concerning problem identification as follows:

> "The start of research is the most important and difficult stage of research. It requires far more than logic; it includes procedures which cannot be neatly categorized and communicated." [Hildreth and Castle, p. 38].

There are several characteristics which a problem statement will possess if it defines a researchable problem within the context of applied research. These characteristics are:

1. Problems reflect felt needs,
2. Problems are non-hypothetical,
3. Problems suggest meaningful and testable hypotheses,
4. Problems are relevant and manageable, and
5. A researchable problem differs from a problematic situation.

Problems Reflect Felt Needs

A problem exists when there is a need felt by a client. This client may be an individual, a group, or a society. Distinguishing who is the client is important as a felt need at the individual level may be biological (a want that must be satisfied for survival) or psychological (a desire to achieve some goal). A group may be dealing with a sociological goal for which various functional requirements figure into a need framework that includes both biological and psychological concerns.

---

[3]For a discussion of various attitudes toward problem identification in research see Brauer and Hunter; Johnson; Manley; Rose; Schermerhorn; Smith.

> A problem exists when there is a need felt by a client.

The need must be "felt" in the sense that the originating party believes that change can be realized, and it may arise from social tensions, doubts, conflicts, failure to realize goals, concern for an anticipated occurrence which might be preventable, or the simple lack of knowledge if this knowledge is necessary to contribute directly to the resolution of another felt need. To be appropriate as a researchable problem, the need must be amenable to change as a result of the information supplied by the research process. Hence, all felt needs are not necessarily operational so far as permitting the formulation of a researchable problem statement.

Problems Are Non-hypothetical

A researchable problem statement must be based on factual evidence. The relationships expressed can be neither hypothetical nor subject to doubt or question in the mind of the researcher. Researchers must use their judgement as they sift through available information to determine to their satisfaction what are and what are not acceptable facts and factual relationships. In a probability sense, of course, few facts can be accepted with complete confidence. However, the facts or factual relationships accepted by the client and the researcher in the problem statement must be such that testing of their validity is unnecessary. If a particular proposition cannot be accepted as fact, then if it is relevant to the case, it must be relegated to the status of a hypothesis.

All researchers and their clients will not accept the same information as facts or factual relationships because individual judgements, knowledge and experience affect this choice [Boulding]. In defining the researchable problem related to a shortage of manufactured dairy products, one researcher, because of her experience and general knowledge, may be willing to accept the proposition that there is no shortage of milk processing equipment in a particular area. For her this forms part of the problem statement. Another researcher may feel that the validity of this proposition is not clear, so in the second case, it must become one of the hypotheses (if it is relevant to the problem). Obviously, the nature of the research project will be different in each case.

This is the stage of the research process in which a review of literature is most productive. By means of this review the researcher seeks to utilize the experience of others to help in the orientation of the project. It should be apparent that this kind of orientation can change the focus of the research effort and can have a significant impact on the productivity of the research resources.

Problems Suggest Meaningful, Testable Hypotheses

Because the statement of the problem serves to orient the entire research process, it must suggest testable hypothetical relationships. Hypotheses are formulated as partial explanations of the unknown relationships which create the problem, and those which cannot be tested will be of little assistance in the resolution of the problem. Hypotheses are testable when information about their validity may be collected and analyzed.

The hypotheses must be developed from the problem statement in a manner which does not result in trivial solutions. Triviality indicates a tautology, an obvious solution, or an infeasible solution. The hypothesis that "per capita consumption of food products is low because there are too many people," derived from a problem statement referring to the existence of hunger in a country, is such an hypothesis. Even if the hypothesis is substantiated, it results in an answer which is of little or no use in the alleviation of the immediate felt need.

Failure of the problem statement to suggest testable hypotheses for resolution of the problem under investigation may signal several weaknesses in formulation of the problem. The scale of observation used relative to a causality may overstate or understate the need expressed by the client. Insect damage to a crop may significantly reduce yields and producer incomes yet high yields, resulting from no insect damage, and high production may also reduce incomes due to a market glut and low prices. But if we are not sure which causalities are important because the problem is not adequately specified, then hypotheses to give direction to the research cannot be readily formulated. Similarly, if the specification is either too general or too specific the observed problem may have no causal context worthy of hypothesis formulation. Adjusting the scale of observation (aggregating or disaggregating) may lend meaning to causality. Yet, the psychology of a crowd of farmers relative to an insect infestation cannot be deduced from an individual farmer and vice versa; for some insect damage may lead to starvation. Thus, if useful hypotheses are to result from a problem statement, care must be taken early so that the level and nature of causality coincides with the client's situation.

Problems are Relevant and Manageable

Agricultural scientists tend to work at extremes. "They tend to work either on problems where the outcome is highly predictable but which has little impact on problems or on problems so large as to be unmanageable" [Hildreth and Castle, p. 38].

This comment was made with reference specifically to agricultural economists but it applies to other disciplines as well. An agronomist designing an experiment to determine yield response on a certain soil, even though information about similar soils is readily available, is an example. The researcher knows with a high degree of certainty whether or not a response will be observed.

Over ambition, lack of adequate forethought, and inexperience are the principal causes of unmanageable projects. An over-ambitious researcher may be motivated by a desire to study all the problems in a given sector so that he can answer any question that arises. Or an unmanageable research project can result from suggestions by clients or administrators unfamiliar with the discipline of the research. In this case the researcher should not accept the project without first more precisely defining the problem and reducing the proposed project to manageable proportions within the time requirement and within the human, financial and other resources available for the project.

> An unmanageable research project yields few benefits to anyone.

An unmanageable research project yields few benefits to anyone. The most likely result of such a project is superficial treatment of part of the problem and the neglect of other parts. Little detail will be achieved, and much that is presented may be found to be inaccurate or insufficient. The contribution of this type of study will almost always be less than that of a more specific study which examines fewer phenomena, but does so in more detail.

Researchable Problems vs. Problematic Situations

Confusion is likely to exist regarding the differences between a problematic situation and a researchable problem. In our context, there is a real and functional difference between the two. First, a problematic situation is a phenomenon which exists; a researchable problem must be identified and defined. A problematic situation represents a generalized situation but a researchable problem, expressed in the above terms, must be specific. "An increasing rate of crime in the cities," assuming the statement is true, is an example of a problematic situation -- it exists and it pertains to felt needs. But in no way can the statement be construed as a statement of a researchable problem. The expression of a problematic situation can serve as general orientation to the broad area of interest of a research project, but it is not sufficient to serve as a guide to a research project.

A problematic situation can be the source of a number of researchable problems. Different clients, different research administrators and different researchers are likely to arrive at a variety of researchable problems from the same problematic situation. That is to say there is no single right, or correct, researchable problem -- there is only a correct framework within which the researchable problem must be expressed to be useful to the researcher. The selection of the researchable problem which will actually be studied will depend on the situation within which the research is being conducted.

Examples of Problem Statements

In evaluating statements which are considered by their author to delineate researchable problems, it is difficult, if not impossible, to determine when deficiencies are due to lack of care in problem formulation on the part of the researcher or to an unclear concept of the nature of the problem. An example of such a statement is the following:[4]

---

[4]Many examples used in the remainder of this chapter are taken from class papers and research proposals submitted by graduate students in the Department of Agricultural Economics, Instituto Colombiano Agropecuario, Bogotá, Colombia, in the late 1960s.

1. "Rural unemployment created by an increase in the use of agricultural machinery."

There is no doubt that the author was focusing on a problematic situation and this interest had to do with unemployment. But it is not clear what specific problem existed in his mind. The statement does not provide the basis for a research project though it might spark a lively debate. As it stands, it does not meet the basic requisites for the specification of a researchable problem. Other examples of incomplete or incorrectly formulated problem statements are the following:

2. "Few cattlemen vaccinate against hoof and mouth disease."
3. "Exportation of fresh meat from Colombia."

Let us analyze the above three statements from the point of view of the requisites necessary for a correctly specified problem statement. Do they concern a felt need? Probably all do, but what exactly is that felt need? The felt needs suggested by those statements are numerous. Are the rural unemployed a problem from the standpoint of crime or poverty in the cities, or is the author perhaps considering those unemployed as a source of inexpensive labor for rural industry? Is the author (or client) of the second statement the president of a drug company or is he the head of a meat export company? Is the third statement a problem from the point of view of Colombia, which may be looking for increased foreign exchange, or perhaps of Argentina which may be considering Colombian competition in the world market? Each of these points of view suggests a different problem, and hence a different research project. It would be folly to initiate a research project on any of the above topics without more complete specification of the client's felt need.

Are the statements hypothetical or subject to question in the mind of the author? This is the most difficult test for a third person to make because he or she is not aware of the precise knowledge held by the researcher or client. For this reason, proper background information in the research project statement is critical. Not everyone will automatically be convinced that the increased use of agricultural machinery leads to rural unemployment. However, if such a relationship is acceptable to both the client and the researcher, and is considered firm and non-hypothetical, it can be acceptable in the problem statement as a non-hypothetical relationship. The second and third statements do not express causal relationships and hence are not subject to this criterion. They are simply incomplete.

Do the statements suggest testable hypotheses? The first statement could be formulated as a testable hypothesis, but if it is to be accepted as a factual relationship for a problem statement, then it cannot be a hypothesis for the purposes of the research. Although testable hypotheses associated with the other two statements could be contrived, any specific relationship between them and the problem in the mind of the reader would be purely coincidental.

Another problem statement will be analyzed to determine if, or how, it might be improved.

4. "Deficient milk production reduces domestic consumption and makes imports necessary in this agricultural subsector."

On the surface this appears to be a more complete problem statement than any of the preceding. Certainly it represents a felt need; in fact more than one. Is the problem to which the author is directing herself a production problem, one of poor nutrition because of inadequate quantities of milk, or is she concerned with the expenditure of scarce foreign exchange? From the present statement, it is impossible to be sure.

The relationships expressed probably meet the non-hypothetical requisite, though not necessarily. The first relationship is really a tautology if "deficient" is defined in terms of domestic consumption. The second relationship could be hypothetical unless government policy, for example, has provided for the imports of milk to make up deficits in domestic consumption.

The statement does not legitimately suggest any hypotheses except the trivial ones that increased production would increase consumption and/or reduce the necessity for imports. And these are not really testable in the usual research framework. Hence, even though the problem would be amenable to one or more meaningful solutions, it is probably not researchable within the usual limits of time and funds. Several years would be required to determine experimentally if increased domestic production would, in fact, result in increased consumption.

In order to improve this statement, it is first necessary to focus more precisely on the orientation of the author with respect to her client's felt need. It turns out that the author of the statement was concerned with problems of production, principally with high costs, low productivity, and the small profit margins of the producers of milk. Her research interest then focuses on the dairy herd and the dairy farm, and possibly on the farm-to-market process. Although it will never be possible to determine the exact nature of a proposed research project from the problem statement alone, a more precise statement is necessary before one can proceed to formulate hypotheses and specify objectives. A better formulation of the above statement resulted from more reflection by the author on her orientation:

5. "The low level of technology on dairy farms contributes to high costs of production, low average productivity, and a deficient marketing system for milk. These factors cause low profits to the producer, price fluctuations for the consumer, and a deterioration in the balance of payments because of the necessity to import milk."

In this form, the statement implies, in accordance with the requisites, that the author accepts as fact that a low level of technology is a causal factor in low productivity, high costs of production and a deficient marketing system for the product. Furthermore, she accepts that these factors cause low profits, price fluctuations and the necessity of importing milk. By accepting these relationships as fact, they cannot appear as hypotheses to be tested by the research process. If the researcher or her client cannot accept any of these relationships as factual, then the problem statement should be modified and the doubtful statements submitted as hypotheses to be tested (assuming they can be tested within the limits of the available research resources).

Formulation of the Hypotheses

It is clear that the logical sequence of events in the process of applied scientific inquiry begins with the observation of phenomena in the empirical world in common sense terms. Determination and classification of these events into problematic situations and specific researchable problems set the stage for postulating various potential means of problem resolution. This part of the process involves the formulation of hypotheses which formalize the premises to be tested by the research effort.

Hypotheses are tentative propositions that must relate to the problem so as to be helpful in providing means of resolution. Any suggested or indicated means of resolving a problem must be formulated so as to be testable and the relationship to the problem must be evident. To be logically complete and functional a hypothesis must involve a relationship. Such a formulation constitutes, implicitly if not explicitly, a hypothesis in the form of "if-then" propositions. The "if" clause describes the relationship between the postulated condition and the proposed result. For example, "If Colombia can increase its beef production by 15 percent and reduce the cost of production by 10 percent, then it can successfully export beef and meet domestic requirements." The first clause, "If Colombia can increase its beef production by 15 percent and reduce the cost of production by 10 percent," sets the conditions that must be met and the remaining clause relates the proposed results. "If farmers adopt a disease resistant bean variety, then costs of and chemical use for control can be reduced and yields will remain the same" is another hypothesis in which the postulated condition and the proposed results are evident.

> Hypotheses are tentative propositions that must relate to and help resolve the problem.

Hypotheses are derived from the observations and relationships accepted as, or assumed to be, fact in the problem statement. Theory and assumption complement the problem statement as the process begins to shape the research activity from the problematic situation. They provide the guidelines for the type of data and techniques necessary for analysis. This implies that hypotheses are formulated before the data collection activity of the research project has started. In this sense hypotheses indicate the direction for data collection; hypotheses that are formulated to explain observations after they are collected may not be useful for problem resolution. The hypotheses, then, are a necessary link between the problem, the data collection and the analytical stages of the research.

The basis for correct formulation of hypotheses is the knowledge of the researcher, this knowledge being founded in theory[5] and agumented by the researcher's experience. The broader the experience of the researchers in relating theory to applied problems, the more efficient they will be in formulating appropriate hypotheses. Whereas the researcher

---

[5]For this argument and others relating theory to the research process see Allin; Christensen; Halter; Kuznets; Parsons; and Chapter 2.

and the client, jointly, must share the responsibility for problem specification, it is primarily the responsibility of the researcher, as the expert in the field, to formulate the hypotheses.

Characteristics of Hypotheses

Hypotheses appropriate to applied research have the following characteristics:[6]

1. They must be formed as "if-then" relationships or amenable to this formulation, and stated in such a manner that their implications and relationships to the problem can be shown logically. The explicit use of the words "if-then" is not necessarily required; the relationship, however, is critical and often an explicit "if-then" statement will assure an accurate interpretation of the relationship.
2. They should be stated as simply as possible both in terms of theoretical complexities and implications and in terms of number of variables.
3. They must be capable of verification or rejection within the limits of the research resources.
4. They must be stated in a manner which provides direction for the research. The hypotheses, when well formulated, will suggest the appropriate data and analytical techniques for testing that should be employed in the research process. Thus, a set of hypotheses can be thought of as a plan of action.
5. Taken together, they must be adequate and efficient in suggesting one or more meaningful solutions to the problem. They must provide for an acceptable level of confidence in the results, but at the same time economize in the use of scarce research resources.

---

[6]Note that there are three categories of hypotheses: 1) the null hypothesis, H_0, for statistical testing (see Steele & Torrie), 2) those contributing to generalizations, laws and theories (see Chapter 3), and 3) those that identify causality for problem solving relationships in applied research. Here we discuss the third category.

Some Examples of Hypotheses

To continue with the proposed dairy study, the following are the hypotheses which were submitted in the second draft of the proposal:

1. "Increasing the level of technology and of physical production in association with a reduction in the costs of production, would result in stability between supply and demand."
2. "Providing more financial resources for increasing production and restructuring the market channels would allow simpler price regulation."
3. "Establishing milk regulations would provide optimum quality and a price warranted by that quality."

Although all the hypotheses are generally related to the problem as stated, it is quite obvious that they encompass a much broader front than we were imagining from the reformulation of the problem statement. This revelation leads one immediately to the suspicion that the hypotheses violate the second requisite of hypotheses -- simplicity. The first hypothesis encompasses a fair proportion of all economic theory and certainly does not clearly specify how one moves from supply to demand with ease. Any reasonably low-cost means of testing this hypothesis is hard to imagine. The second and third hypotheses are apparently related to the problem through the concern with price fluctuations and, perhaps with the low profits of the producer. However, the relationship to the problem is only tenuous at best, and actually introduces new concepts that were not in the problem statement. Rather than helping to clarify the research proposal, as they should, these hypotheses tend to add confusion. Clearly, they do not provide direction or guidance for the research which is one of the primary functions of hypotheses.

Because the author of the hypotheses in the example did not progress beyond this point, we must now begin to act as if we were the researcher and formulate the hypotheses according to our understanding of the problem. In doing so, it may be necessary to better identify the problem itself (normally done in consultation with the client) so that the proposal can be improved, but because this occurs many times in the formulation of a good research proposal, we will not be wasting effort. Few people are capable of writing an acceptable research proposal on the first attempt. Several modifications usually are necessary as the orientation and focus become clearer. The final version of the problem statement in the previous section read as follows:

> "The low level of technology on dairy farms contributes to high costs of production, low average productivity, and a deficient marketing system for milk. These factors cause low profits to the producer, price fluctuations for the consumer, and a deterioration in the balance of payments because of the necessity to import milk."

> Few people write an acceptable research proposal on the first attempt.

Let us assume that we were correct in deciding that the orientation of the proposal was more toward production than toward marketing. Let us also accept the relationships expressed in the problem statement although we may not be comfortable with some of them. Then, presumably, the focus of the proposed research project is the low level of technology on dairy farms, this being the cause of the unfavorable relationships expressed in the problem statement.

Because the hypotheses must be testable within the resource limitations, care must be exercised in comparing resource requirements and limitations before continuing with the formulation of the proposal. Assume that the client in this case is the Planning Department of the Ministry of Agriculture. In preparing their plans for the following year, they are considering the long range necessity of importing milk. They have requested a series of studies to help them clarify the situation so they can make firmer predictions and to aid the Ministry in establishing domestic policies. The Planning Department staff consider that the country is capable of producing more milk but they are not sure why it is not doing so. One factor that is evident is that there is a low level of technology used in the dairy industry. In this particular study their desire is to determine why technology has not improved in recent years, because general knowledge far exceeds implementation. Professionals will be provided to work full time on the project, and the Ministry wants a preliminary report in four months and a final report in six. Generally, personal computers are available and it will be possible to hire the services of some interviewers for a short period of time if necessary.

The first hypothesis might well treat the profitability of the new technology because, if it is not profitable, most assuredly the producers will not adopt it. For a beginning, let us use:

1. "There exist in the country improved technologies for dairy production which, if used by producers, would increase their profits."

Stated in this fashion the hypothesis does indicate the possible direction which the research might take. In verifying or rejecting it, one method might be a simple review of literature to determine if we can satisfy ourselves as to the existence of profitable new technologies. Perhaps the technologies which are available have never been subjected to tests of profitability, which would indicate the need for some partial budgeting or possibly linear programming for a series of typical farm resource situations. Of course it might also be true that the client is satisfied that there are profitable new technologies and therefore does not desire or require verification of this hypothesis. Assuming that we will need to include this hypothesis in the project, further clarification will be made in the statement of the objectives and the procedures.

The hypothesis can be stated in an "if-then" form, "If farmers adopt presently known technology then their profits will increase," and the relationship to the problem statement is clear. Hence, it satisfies the first requisite of hypotheses.

Depending on the other hypotheses which are derived, all of the previously discussed means of verification or rejection of the hypothesis could be accomplished under the limitations of the budget which we have described.

The theoretical implications of the hypothesis are about as simple as it is possible to make them and still retain meaningful relationships to the problem. To be profitable, the new technologies must fit within the resource restrictions of the farmers and must be profitable given present or projected price situations. Notice that it is not adequate, even though it is simpler, to hypothesize only that the new technologies could increase farmers' production.

The last criterion of hypotheses applies to all of them taken together, so it cannot be applied individually except in the sense that it would not be adequate to consider only an increase in production without considering, at the same time, the profitability of the practice.

Assuming that in the course of the research we will be able to accept the first hypothesis, we must then consider additional hypotheses because the first, alone, is not adequate to answer the question put to us by the client. The second hypothesis might be that:

2. "Farmers are not aware of the new technologies."

This is stated as a fact, but the "if-then" implication is that if farmers are not aware of profitable alternatives then they will not (or cannot) adopt them. The direction of the research in this case is also indicated. A sample survey of farmers should be able to provide the evidence necessary to accept or reject the hypothesis.

Another hypothesis could treat credit, or the related financial situation of the farmers.

3. "If special credit sources are not made available, farmers will not be able to adopt the new technologies."

This can be considered a broad hypothesis with two sub-hypotheses which will be tested,

a. "Farmers are unable to adopt new technologies because internal financial resources are limited."

and

b. "Farmers are unable to obtain credit, which limits their ability to finance changes in their production technologies."

An additional hypothesis could treat the problem of price instability from the point of view of the producer.

4. "A milk price stabilization program could induce farmers to adopt improved production technologies."

This hypothesis is not quite as straightforward as the others. To be able to verify or reject it empirically, a great deal of time and money would be involved -- certainly more of both than we have available within our research budget. However, attitudes of farmers toward a price program could be ascertained and certain conclusions drawn regarding the probable outcome of such a program. If this hypothesis is included, the precise nature of the research process will need to be specified in the objectives and the procedures, so the client is not mislead.

Other possible hypotheses could be suggested, but those proposed above are adequate for present purposes. Probably all can be accomplished within the research budget, but it will be necessary to clarify them further (and perhaps modify them) as we develop our statement of objectives and procedures.

In summary, our tentative hypotheses are the following:

1. There exist in the country improved technologies for dairy production which, if used by the producers, would increase their profits.
2. Farmers have not adopted the new technologies because they are unaware of their existence.
3. Special credit sources are necessary if dairy farmers are to adopt improved technologies of milk production.
    a. Dairy farmers are unable to adopt new technologies due to internal financial resource limitations.
    b. Dairy farmers are unable to obtain credit, which limits their ability to finance changes in production technologies.
4. A milk price stabilization program could induce farmers to adopt improved production technologies.

Are these hypotheses, taken together, adequate and efficient in suggesting a means to one or more meaningful solutions to the problem? If they cover the range of possibilities open to the government (extension programs, credit programs, and/or price programs, for example) they should be adequate in suggesting guidelines to one or more meaningful solutions. They are efficient if we cannot contrive other hypotheses which could provide solutions to the problem with the use of fewer of our research resources, do so in less time, or result in more precise information within the research budget.

Delineation of the Objectives

Objectives usually are expressed in lay terminology and are directed as much to the client as to the researcher. The primary objective of applied research will be either (1) to suggest or recommend to the client practical means of problem resolution, or (2) to provide information to clarify an unknown situation. Generally, the objectives taken as a group will (1) define the limits of the research project for the researcher, (2) clarify the means of conducting the research, (3) identify the client or clients, and (4) describe the expected product of the research for the client.

> Objectives specify what the researcher intends to do and what information will be obtained for the client.

The objectives link the theoretical relationships presented or implied in the hypotheses to the analytical and methodological orientation necessary for conducting the research. An objective specifies what the researcher intends to do or find in the project and suggests one or more research procedures to be used. Later in the research proposal the specific procedures must be defined. These procedures of course, must be appropriate, not only to the stated objectives but also to the resource availability as discussed in Chapter 4. It is necessary, therefore, that the objectives be sufficiently broad to satisfy the needs of the client but also sufficiently specific to conform to budgetary restrictions.

An erroneous idea of the nature of the objectives of a research project should be clarified. Research objectives are neither political objectives nor are they objectives of an action program of the government. Objectives of a research project suggest what information will be obtained for the client to help resolve the problem which initiated the research. This information, in turn, can be used by the client to establish an action program. For example, "to redistribute land in Mexico" suggests a government action program. This cannot be an objective of a research project. But, "to study the past and present agrarian reform programs in Mexico and their effects on land redistribution" is a legitimate research objective related to a felt need concerning land use and productivity. In the terms of our dairy problem, "to establish a credit program for dairy farmers" is not a legitimate research objective, though it may be a feasible solution in the mind of the researcher and the client. "To determine if dairy farmers are able to obtain credit for improving methods of production," or "to recommend methods of resolving credit deficiencies if identified," on the other hand, are acceptable research objectives.

Given these conditions, the primary objective of the dairy study which we are proposing could be to:

Determine the obstacles to the adoption and optimal use of new technology by dairy farmers. This objective precisely describes the purposes of the project. Accomplishing this objective will provide information to clarify an unknown situation for the client, and, if we are successful should provide the information needed to make policy decisions and predictions concerning future milk supplies.

Secondary objectives can include the following, listed in order of their relationship to the hypotheses:

1. Determine if presently known modern technology is profitable to the dairy farmers given their present resource situation and market outlet.
2. Determine if the Extension Service is effectively providing necessary information to farmers concerning possible alternatives for production.
3. Determine if the required changes to adopt new technologies on farms are outside the financial means of the farmers.
4. Ascertain whether present credit sources are adequate in providing for the farmers' needs related to new practices.
5. Obtain farmers' opinions about a price stabilization program and their possible reaction to it with respect to changes in use of technology and concurrent production practices.

Notice that each of these objectives is directly related to a hypothesis and helps to clarify the direction of the research. If the client does not understand the terminology used in the hypotheses, he should be able to understand the nature of the research from reading the objectives. Also, the objectives clarify the means of conducting the research. It is clear that interviews with dairy farmers will be necessary and, in general, what the content of these interviews will be. Specific questions must be written from the guidelines given in the objectives. These questions either singly or as groups will provide the information to test the hypotheses.

Of special interest is the fifth objective. The related hypothesis (a milk price stabilization program could induce farmers to adopt improved methods of production) states a positive relationship without giving any indication of the nature of the test which the researcher has in mind. The client in such a case could well formulate an erroneous impression of the nature of the results which the researcher proposes to present. The fifth objective specifically states that the researcher expects no more than to obtain the farmers' opinions and possible reactions. This is clearly different from empirical evidence which the client might otherwise expect if he only had the hypothesis for information.

We have been talking about a special or explicit client but other clientele could also benefit from the research in question. If the researcher is hired specifically by a client, and the client considers the research private, then, of course, this client is the only one who will directly benefit from the work. However, as is more often the case, this type of research is undertaken by a public or semi-public entity so the research becomes public property and the identification of a broader audience is useful. In the context of the present problem, the full clientele could be identified in a sixth objective:

6. Provide information to farmers about the profitability of new production technologies, to bankers and other credit institutions of possible sources of new business, and to government planners to aid them in making decisions concerning the future of the dairy industry.

Summary

Three of the most important and critical aspects of the planning of a research proposal -- problems, hypotheses and objectives -- have been presented and discussed. These three parts are not independent from each other, nor are they independent from other portions of the proposal and aspects of conducting the research to be discussed in following chapters. They serve as a framework for developing the data collection and analytical procedures, the budget including time sequences, and the publication plans for the research results.

> Careful development of the problem statement, the hypotheses and the objectives is the key to efficient research and the most productive use of time by the researcher.

Time spent in careful development of the problem statement, the hypotheses, and the objectives is the key to efficient research and can well be the most productive use of time by the researcher. Even in cases where the researcher may have only one month, one week, or perhaps just one day to provide an answer, the time spent in this phase of the research is critical to the success of the undertaking. On many occasions when a person is given a rush task, the tendency is to "come up with something." Little time is devoted to analyzing the situation to determine precisely what the client wants, what is really needed, and what resources are available to accomplish the necessary task. Too often the results have no value because the "something" which the researcher "comes up with" is not really related to the problem of the client. Recall the Minister's frustration with her researcher in the opening dialogue.

Problems appropriately specified for applied research have the following characteristics:

1. They are based on felt needs of individuals, groups, and societies;
2. The causal relationships expressed in a problem statement are not hypothetical and are relevant to the problem;
3. Problem statements must suggest testable hypothetical relationships that, when analyzed, yield relevant and non-trivial results; and
4. The problem and the research to resolve the problem must be relevant and manageable within resource restrictions.

Researchable problems can be distinguished from problematic situations in that numerous researchable problems can be formulated from a problematic situation.

The hypotheses serve as guides to executing the research. Hypotheses must:

1. Be stated to provide direction for the research;
2. Be formulated as causal relationships with "If: Then" implications;
3. Be capable of tests within the limits of the research resources;
4. Be stated as simply as possible; and
5. As a group be adequate and efficient in suggesting means to one or more meaningful solutions to the problem.

Objectives in general describe what is expected to be achieved by the project. Specifically, objectives:

1. Define the limits of the research project;
2. Suggest or clarify the means of conducting research;

3. Describe the nature of the potential research product to the client; and
4. Identify the client or clients.

The following problem statement, hypotheses and objectives are those of the example discussed in the chapter.[7]

Problem:

The low level of technology on dairy farms contributes to high costs of production, low average productivity, and a deficient marketing system for milk. These factors cause low profits to the producer, price fluctuations for the consumer, and a deterioration in the balance of payments because of the necessity to import milk.

Hypotheses:

1. There exist in the country improved technologies for dairy production which, if used by the producers, would increase their profits.
2. Farmers have not adopted new technologies because they are unaware of their existence.
3. Special credit sources are necessary if farmers are to adopt improved production technologies.
   a. Farmers are unable to adopt new technologies due to internal financial resource limitations.
   b. Farmers are unable to obtain credit, which limits their ability to finance changes in production technologies.
4. A milk price stabilization program could induce farmers to adopt improved production technologies.

Objectives:

The primary objective is to:

Determine the obstacles to the adoption of new technology by dairy farmers. Secondary objectives are:

---

[7]See the appendix for examples of complete project statements.

1. Determine if presently known modern technology is profitable to the dairy farmers given their present resource situation and market outlook.
2. Determine if the Extension Service is effectively providing necessary information to farmers concerning possible alternatives for production.
3. Determine if the required changes to adopt new technologies on farms are outside the financial means of the farmers.
4. Ascertain whether present credit sources are adequate in providing for the farmers needs related to new practices.
5. Obtain farmers opinions about a price stabilization program and their possible reaction to it with respect to changes in use of technology and concurrent production.
6. Provide information to farmers about the profitability of new production technologies, to bankers and other credit institutions of possible sources of new business, and to government planners to aid them in making decisions concerning the future of the dairy industry.

As a summary of this dairy example and as a means for considering the process of forming and prioritizing hypotheses, a tree diagram is useful. Many hypotheses can be specified. Some definitely should be tested before others to achieve research efficiency; some may be too costly given available research resources; others may be easily testable but not high enough in priority to merit even minor expenditure; and still other relationships may be sufficiently obvious that the use of theory, assumption or past experience will adequately substitute while maintaining applied research quality. The tree diagram in Figure 6.2 can be viewed as related cause and result situations, each selected to further specify the problem for purposes of determining possible explanations and solutions. Presentation of the research proposed in a simple tree diagram is an excellent means for communicating with the client or initiator of the search to solve the problem.

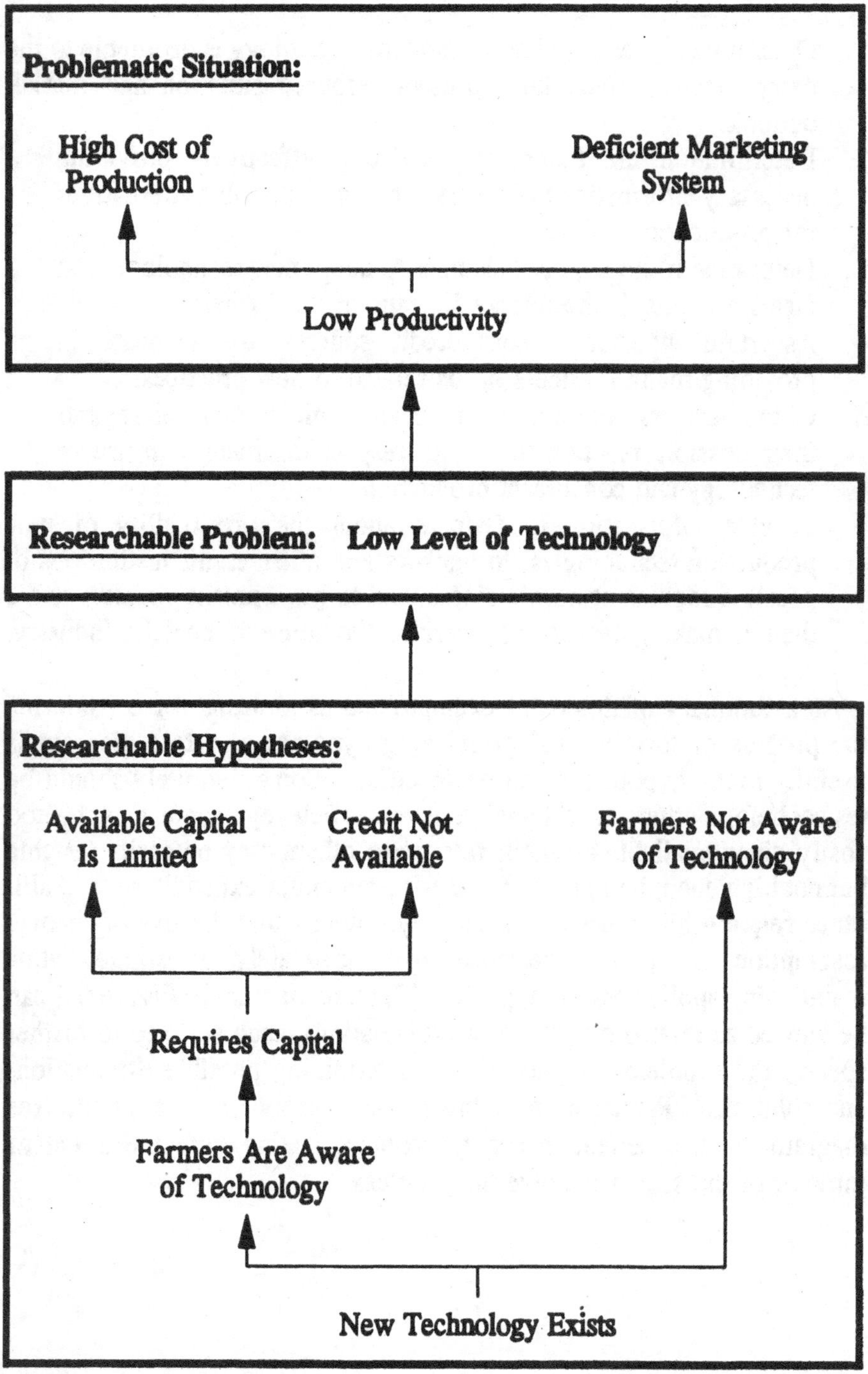

FIGURE 6.2 Tree diagram of dairy technology example.

PART THREE

Conducting Applied Research

7

The Orientation to Observation

Research is a holistic endeavor that requires a full range of information gathering and analysis capabilities. Decisions about the approach to be taken for information gathering require knowledge about the problematic situation to be researched.

The execution of research necessitates some orientation to observation. That is, what will be the orientation to the collection of quantitative and qualitative information that pertains to the project? How much objectivity must be maintained in the process of observation? These are important questions because utilization of the resulting information provides the basis for testing the hypotheses and reaching conclusions useful in the resolution of the problem which initiated the research. From these conclusions appropriate recommendations are expected by the client. If the project has been properly planned, the type of data which are needed have been determined prior to conducting the research. The researcher will know if only secondary sources will be used or if primary information collection will be necessary. The planning phase of the project will also determine if experimentation is necessary or if the source of data will be non-experimental, and will consider the degree of desired or potential interaction of data from each source. Experience with and availability of appropriate information and data collection procedures influence the degree of success in problem oriented research. Experimental research by bio-physical scientists and survey research by socio-economic scientists are commonly used for agricultural investigations. Neither is exclusive of the other. Both assist with problem identification and both incorporate some of the primarily social science tools of either rapid or prolonged information gathering. Many of the techniques share some common practices and principles. Practitioners of one technique often gain by exposure to others.

This chapter attempts to structure the orientation to research observation around approaches noted as experimental/non-experimental, human subject interaction/non-human subject interaction, and descriptive/predictive to prescriptive. These distinctions are dealt with because they represent some of our methodological vocabulary and, while often serving to confuse communication, seem to influence research planning, management and policy.

Experimental and Non-experimental

Experimentation has been the principle basis for obtaining scientific information and will continue to remain of paramount importance. The use of non-experimental data, however, is becoming more prominent, due to improvements in measurement techniques and development of more adequate data series. The advantage of experimental data over non-experimental data is basically the degree of control which the researcher is able to exert over the variables included in the study. In most experiments, with the use of the proper design, the researcher can select the factors which will vary, the levels at which they will be included in the experiment, and the pattern in which they will be used. By using appropriate equipment he or she can also usually obtain quite accurate measures of the input variables (which are assumed to be precise) as well as of the results of the experiment. With non-experimental data, the levels and combinations of variables are predetermined by nature or society, so the researcher must measure and use them as they exist. Because many of these variables are very difficult to identify and measure, non-experimental data are usually more subject to imprecision than are experimental data.

> The advantage of experimental data over non-experimental data is the degree of control held by the researcher over the variables under study.

A major difference between experimental and non-experimental research, as stated above, is the degree of control the researcher exercises over the variables being studied or measured. In an experiment, the researcher controls the design and levels of certain variables and the measurement of phenomena resulting from the experiment. In non-experimental research, the researcher in most instances cannot determine the design and level of the variables nor directly measure the phenomena, but controls only the technique used in measurement. For nonexperimental observations, sample survey design plays a role comparable to that of experimental design when experiments provide the observations to be analyzed.

Non-experimental researchers rely upon interviews, observation techniques, discussion, and formal questionnaires to communicate their measurement needs to informants or respondents. Often it is the informants or respondents, not the interviewers, who make the measurements requested in the interview. The informants' or respondents' experiences may be available in farm records or household budgets, but frequently their responses are based upon a subjective evaluation of the phenomenon as they best remember or can recall it.

Thus, in non-experimental research the researcher usually controls only the general levels of variables through: stratification and sample selection of informants or respondents; specification of the points to be discussed, questions to be asked and questionnaire development process; and training of research assistants and interviewers. Successful non-experimental measurement rests primarily upon all these activities. But, for example, even with a good sample, a tested questionnaire, and a well trained interviewer, the researcher cannot completely control interviewer-respondent communication, part of which may be misleading [Good and Hatt]. Several approaches to improving the quality of communication are being refined to assist in measuring socio-economic and other behaviors.[1]

Degree of Subject Interaction

In all instances the researcher influences the research process and outcome. Human subject interaction, either as the unit of study or as the person responsible for the study, varies greatly depending upon the approach utilized. We consider six combinations of experimental/non-experimental and behavioral/bio-physical research orientations relative to the degree of human subject interaction (Figure 7.1). There is a

[1]See Dillon, Morgan, Krueger, Reason, Yin and Whyte.

tendency to increased human subject interaction when moving through the research approaches from bio-physical to behavioral.

Bio-physical

Least human interaction with the research process comes with non-experimental and non-behavioral research on bio-physical phenomena. Such information gathering and analysis usually deals with observation

	Research Approach	Human Subject Interaction (more to less)
A	Bio-physical	X
B	Bio-physical, Experimental	X
C	Bio-physical, Experimental, Behavioral	X
D	Behavioral, Experimental	X
E	Behavioral, Bio-physical	X
F	Behavioral	X

FIGURE 7.1 Degree of human subject interaction in applied agricultural research

and measurement of various natural phenomena. For agricultural research, agro-climatic information is important as a basis for establishing baseline information about conditions that will affect both natural and socio-cultural responses. Rainfall, temperature, day length, altitude, initial soil fertility, ground water quality and quantity, prevailing winds, seasonal variations in natural conditions, actuarial and historical information concerning natural events and many related details generally involve the human agent as an observing or measuring investigator. Measurement itself involves human judgement and observational design but in no way can or should the human influence the natural phenomena to be measured.

Bio-physical, Experimental

Biological and/or physical science through experimentation is sometimes termed "hard science" because of the degree of control exercised over the studied variables. The human agent is removed except as the initiator of the trials or the experiments. Human subject interaction is present in design, management and analysis but is not the subject of study. The range of experimental research opportunities is unlimited as are the techniques used in design, implementation and analysis of the experiments. Hard scientists may work in laboratories (bench scientists) where the natural environment is tightly controlled or in experimental locations where the natural environment is manipulated to a lesser extent.

One common bio-physical/experimental form of agricultural research for example, is the fertilizer trial. Test plots are established and randomly assigned along with control plots to an area of land. Bio-physical responses and interactions, in the form of plant growth and yield, nutrient loss and other measures, are observed for different fertilizer applications. Measurement criteria attempt to isolate and fully remove any human subject interaction and give "scientific purity" to the results.

The choice and application of "hard science" measures and analytical techniques, however, can considerably restrict the potential for immediate use of scientific results. For this reason, various forms of "soft science," non-experimental, or behavioral research are used. These reduce "scientific purity" but enhance adaptation and use.

Bio-physical, Experimental, Behavioral

> Human subject interaction in research varies depending on the degrees to which bio-physical, experimental and behavioral approaches are combined and utilized.

Experimentation involving comparisons between systems, where management decisions are integral to the phenomena under study, can introduce a behavioral component to bio-physical experimentation. As some experimental research moved from agricultural experiment stations to farmers' fields in the form of researcher-managed trials or extension-managed demonstrations the door opened to consideration of farmer participation. Farming systems research and extension has further provided the methodology for participation of the farmer as both a researcher and a researched human subject (decision maker, laborer and consumer) in conjunction with on-farm experimental trials. The intra-household and management components of the farm as a system that condition the natural environment with human behavior are incorporated into a holistic approach to applied experimental research. Fertilizer research, for example, is considered not only in terms of a bio-physical potential but also in terms of farmer willingness and capability to adopt various new fertility practices. Research methods and techniques are altered to incorporate measures of farmer behavior and practice relative to their natural and socio-economic environment. As a result, on-farm fertilizer trial results more accurately reflect farmer conditions.

Behavioral, Experimental

Behavioral experimental research can be viewed specifically as measuring the choices of items presented in an experimental context where the human response or interaction is carefully isolated. Various consumer panels fall into this context. Maize flour testing in taste tests of tortillas formulated from alternative varieties would be an example. Different types of hand-held hoes for soil preparation might be tested for preference in use by farmers. Panel information requires carefully designed experiments so that reliable data, free as possible from phenomena beyond the desire of the researcher, can be obtained. An attempt is made to approximate the laboratory experiment. Detail may require specific ordering or ranking procedures so that preferences can be used for decision purposes. Other data such as those provided from questionnaires, may be taken to complement panel information. Usually it is desirable to fully characterize the human subjects (consumers and farmers in the examples) so that differences among subjects can be distinguished from differences in the test phenomena (tortillas and hoes).

Behavioral, Bio-physical

Observation and measurement of human behavior in the bio-physical context, without experimentation, often appears in the form of census and other major data sets. Demographic information represents aggregates of human response to various phenomena such as climate, employment opportunity, socio-political conditions, cultural preferences and others. Census data reveal varied living patterns, literacy information tells us something about educational orientation, prices provide insights about market conditions, infant mortality rates may be viewed as indicators of health and nutrition conditions and the list goes on. All represent some form of behavioral and bio-physical response by individuals, groups and societies. Note that each is measured in the "past tense" and often as a time series of an event or situation that has occurred or exists because of past action. Biological and/or physical interactions with human agents are observed in the non-experimental context of past human responses. Reasons for the response may be understood by using both survey and experimental research approaches.

Behavioral

It is not easy to distinguish between some behavioral research activities and behavioral, bio-physical research. The distinction is not necessarily important. Certainly the degree of human subject interaction in behavioral research exceeds the other categories listed here. Survey research from interactive interview, discussion, brainstorming and related techniques provide opinions, preferences, judgements, feelings, needs and similar emotive responses as measures of current human activities. Most often such survey data are used to understand or diagnose contemporary situations. Opinion polls may use census techniques but obtain current situational information. It might be said that the half-life of these data generally is less than that of data taken from the other research approaches. People are given full freedom to change their minds and the human subject is almost completely responsible for the nature of the response. Sound questionnaire design, sampling and collection techniques will influence the aggregate results but generally the degree of control is the most limited of all the research approaches. One can imagine inclusion of an experimental element by some major inducement such as rumors of a financial crash, war or similar catastrophic event. Comparisons are made following major events if baseline data are available to document pre-event status.

Conclusion: Descriptive, Predictive or Prescriptive Research?

Many of the techniques and approaches discussed throughout this book can contribute to descriptive, predictive and prescriptive research. Usually experimental scientists balk at becoming prescriptive. Some behavioral scientists, particularly economists, are often criticized for being too prescriptive. Orientations to normative and empirical indicators become very important in debates about research "quality."

Some of these issues are discussed and referenced in Chapter 3. We believe that no easy lines are drawn between client needs for descriptive or explanatory research, prediction (and ex ante evaluation) and prescription.

> No easy lines are drawn between client needs for descriptive, predictive and prescriptive research.

Most clients want to know what will happen if some change is made and usually they would like to know as much as possible about expectations for the future. They also may expect to receive recommendations for a solution to a problem. This is human nature. Humans expect nature to behave in certain ways over time and usually their expectations are met. Yet they are not sure how they should respond. They want similar certainty concerning the collective interactions of humans and nature so that they can make wise individual responses. Herein lies the basis for problem identification and ultimately hyothesis specification with predictive and prescriptive intentions. The research process to meet these intentions and the research results demand precision, accuracy and efficiency relative to the problem under study. We now turn to these ends in Chapters 8 and 9.

8

Agricultural Research Based in Experimentation

Experimental Design

> Research planning is the basis for choosing an experimental design.

Experimental design, the way the experiment is to be set up, is a critical factor in the generation of experimental data.[1] Several designs may serve a particular need, but there will be one that is more applicable to the problem, better suited to the conditions under which the experiment is going to be conducted, provides required precision and is most efficient in the use of the research resources. The most efficient experimental design will provide the most information directly related to problem resolution for a given set of research resources than any other alternative design.

Selection of treatments, the controlled and measured factors, the levels at which the variables are to be included, and the number of replications to be used in the experiment, or whether to use multiple years or multiple locations is not a simple process. The final choice can be complicated in applied research because the conditions under which the researcher is working are often poor in relation to what would be required for desired precision. Research planning is the basis for choosing an experimental design. The design must be related to the

---

[1]Selected references on experimental design include Baum; Baum, Heady and Blackmore; Cochran - 1963a; Peng.

problem and the methods of analysis to be used and these should have been carefully considered in the planning phase of the project.

In the choice of experimental design, the researcher must explicitly balance four factors: precision, accuracy (very different from precision), relationship (applicability or relevance) to the problem, and efficiency in resource use.

Precision is associated with the statistical level of confidence in the results of an experiment. It is related to the probability that the same results would be achieved if the experiment were repeated. In medical research, it is common to demand confidence levels of 0.01 (indicating that the results should be repeatable 99 times out of 100) or higher. Many agricultural experiments conducted under controlled conditions achieve, or are expected to achieve confidence levels of 0.05. Attempts to increase precision can lead to highly controlled experiments which ultimately do not accurately reflect conditions under which the client operates. This can mean that conclusions from the research cannot be applied, nor relevant recommendations made to farmers.

Accuracy is associated with the appropriateness of the results to the conditions of the clients. The high yield levels common with many experiments cannot be achieved by farmers because they cannot duplicate the kinds of conditions under which the experiments were conducted. This means the results do not accurately reflect what would happen if farmers were to use the same treatments. Therefore, recommendations made from the controlled experiments, even if they are precise, may not be relevant to the client nor amenable to solving the client's problems.

Experiments conducted to help solve clients' problems must bear a *relationship to the problem*. That is, they must be designed to provide the *necessary* information and the information must be *sufficient* for drawing relevant conclusions and making recommendations. If not, the experiment may not be applicable nor relevant, or only partially so.

> Precision, accuracy, relationship to the problem, and efficiency must be considered in the choice of experi-mental design.

Finally, the researcher, in consultation with the clients, must balance precision, accuracy and applicability with the availability of resources to conduct the experiment in order to achieve *efficiency in resource use*.

The most efficient design for a given set of resources will provide more accurate and relevant information with desired precision than any other.

Relationship to the Problem

If the design of the experiment is not properly related to the problem orientation of the project it will not determine relationships useful to problem resolution. An example may best serve to illustrate the point.

A common type of design in fertilizer experimentation is the following, with each treatment being replicated a number of times (frequently four).

Design 1

Treatment Number	N	P	K
	Variables		
1	0	0	0
2	0	0	1
3	1	0	1
4	2	0	1
5	0	1	1
6	1	1	1
7	2	1	1
8	1	1	0

In the project proposal an objective might read something like, "to determine the effect of nitrogen (N), phosphorus (P) and potassium (K) on the production of pangola grass in the Cauca Valley." As we now know, of course, part of the difficulty here is that we do not know what the identified research problem is. But that aside, the objective as stated is vague and is not an adequate guide for designing the experiment. Toward what end is the experiment oriented? Does the researcher want to know simply, "Is there a response to N, P, and K?" Or is there more interest in the magnitude of the response, or even the kind or form of response, or interactions over a range of N, P, and K in various combinations?

Each of these questions may well require the use of a different experimental design, number of treatments or number of replications in order to achieve accuracy and efficiency in arriving at an answer. In some cases, two or more questions can be answered efficiently from one experiment, but in others, the attempt to answer too many questions from a single design may render the experiment useless for answering any question.

Without discussing the theoretical logic for the statements, the following can be said about the relationship of Design I to some of the questions which could be asked. When the researcher is interested in knowing only *if there is a response* to each of the three nutrients, it is not necessary to include all the treatments and levels of N in the design. A design with fewer treatments such as the following could be used and would be more efficient:

Design II

Treatment Number	Variables N	P	K
1	0	0	0
2	1	0	0
3	0	1	0
4	0	0	1

Here, instead of eight treatments there are only four. Using four replications of each treatment, only 16 separate plots are required rather than 32 as would be necessary in the first design if four replications were also used. The choice of the level of each nutrient (1 = 100 kg/ha or 1 = 200 kg/ha, etc.) is important because this design provides information only for the one level chosen.

If the researcher wanted to know the *magnitude of the response* for two different levels of N, it would be necessary to include three different levels (0, 1, 2, for example) of this nutrient. Apparently in the design shown it was desired to know something about the magnitude of response to N for two different levels of P, because the N treatments are repeated for 0 and for 1 P. This question can be answered with the use of treatments 2 through 7 of Design I.

The last treatment of Design I can only answer two questions, (1) "Is there an effect from K when N = 1 and P = 1?" This requires treatments 6 and 8. No other information will be available on the effect of K. The last treatment along with the control (treatment 1) can show the response of N and P together when no K is applied.

Because we are interested in applied research, let's suppose that the ultimate use of the data from the proposed experiment will be to make fertilizer recommendations to farmers in a specifically characterized situation. To what extent does Design I permit us to make the appropriate analyses? One alternative is to compare the cost and return of each treatment with the control (0--0) and determine the net return from each. Assuming that there were significant differences between treatments, we could then recommend the treatment with the highest net return.

Another alternative would be to determine the functional relationship between N and production (production function for N) for each of the two levels of P. Assuming the curves thus generated were of proper form (satisfying the conditions for optimality), we could determine the economic optimum quantity of N for each level of P, make a cost-return estimate for each, and choose the one with the higher net return as that which we would recommend.

> In determining relationships to assist with problem resolution, the design of a *useful* experiment must be related to the problem.

The second procedure would probably result in a somewhat different recommendation than the first method. There is obviously less exactness with respect to N in the first method than in the second but there may be even less exactness with respect to P because we have only two treatments to choose from. Apparently, the design is not well suited to answer the question the farmers might ask, "What is the *best* combination of N, P, and K to use?" even though it does appear to be at least reasonably suitable for answering some other questions.

To answer the farmers' question may require a more complicated design than available resources allow. If there are sufficient resources, much more exactness is possible. As an example, a complete factorial with 3 or 4 levels of each nutrient would require 27 or 64 plots respectively for each replication , but could provide rather exact answers for the farmers (the use of the 4^3 factorial, of course, would provide more exactness than would the 3^3 factorial but it also requires more resources). Another, more efficient design is the rotatable central composite which is a modified factorial that requires only 15 plots for each replication when three nutrients are considered. For this last design, fewer replications are necessary so that with only about 40 plots a complete experiment can be conducted,[2] providing a wide range of information.

While it is true that precision can be achieved with highly controlled research such as that conducted on experiment stations, the results do not necessarily accurately reflect what farmers might expect from applying the treatments on their own farms, under conditions they can create. The difference is attributable to the artificially created research environments on experiment stations versus the economically, socially, politically and/or culturally achievable environments found on farms. Experimental designs to cope with these differences must be used to make research results more accurately reflect the conditions clients can realistically expect. One such approach is based on Modified Stability Analysis[3] which uses research results conducted under variable farm conditions to generate an environmental index to help predict accurate results, realistic for farms. This approach is discussed later in the chapter.

In summary, the problem toward which the research is directed has a strong bearing on the type of experimental design which should be chosen if the researcher is going to use experimentation in the execution of the project. Should the magnitude of the problem require a very detailed experimental design, the researcher should be encouraged to consult a competent statistician when one is available. Also, if the researcher is going to use secondary experimental data, the design which was used in the experiment should be considered when choosing between alternate sources. More also will be said on this aspect in a later section of this chapter.

---

[2]All treatments are not replicated the same number of times [Hildebrand, 1972].

[3]See Chapter 6, page 73 of Hildebrand and Poey, 1985, or Hildebrand, 1984.

Relationship to Resources

The ultimate size and complexity of an experiment depends on many factors including number of independent factors to be controlled or measured, analytical techniques to be used, statistical precision required, quantity and quality of prior information available, the objectives of the research project, the time within which results must be obtained, and the resource constraints encompassing the project.

If it is necessary to determine the best combination of N, P, and K for specific conditions, a rather complex design with a relatively large number of experimental observations (plots) is required. But if it is simply not possible to conduct such a complex experiment, the design must be modified. For example, if some information is available which indicates little or no response to K, then the number of independent variables in the previous example could be reduced to two (N and P) and K could be excluded or held constant at some predetermined level in the experiment. The best combination of N and P can then be determined with a rotatable central composite design requiring only about 26 experimental units rather than the 40 units required for three independent factors (N, P, and K).

The design can be simplified further by considering only one independent factor. This factor could be the one considered most important such as perhaps N or it could be some fixed combination of several factors such as a complete fertilizer containing N, P, and K with an analysis of 10-30-10, for example.

> Size and complexity of an experiment depend on many factors; ultimately time and resource constraints encourage the use of an effective research plan and experimental design.

An experiment requiring only about 15 experimental units will yield as much statistical precision for one factor as 26 for two factors. But it is important to realize that the total amount of information is reduced accordingly as the number of independent factors and resource requirements are reduced. With three factors and 40 plots, we should be able to tell farmers, who are on similar soils to that used for the experiment, and have appropriate resources, how much N, how much P, and how much K should be used to achieve the greatest net returns from fertilizer use. With two independent or variable factors, we have the option of telling them the best combination of only two nutrients and can do that only for some predetermined quantity of the third nutrient. With only one variable factor it is not possible to recommend best combinations but only the best quantity of that single factor. We may be able to recommend 300 kilos per hectare of 10-30-10 as the best quantity of that complete fertilizer, but we would not be at all confident that the resulting 30 kilos of N, 90 of P, and 30 of K would be the best combination of the three nutrients to use.

The effect on resource requirements of the analytical technique to be used is more complex, but an example will serve to illustrate the point. Suppose that it is desired to determine if there are significant differences in the response of three different levels of a new feed additive on fattening steers. A probable design would have three treatments with the additive (one for each level to be tested) and a control. This results in four treatments per replication. Under most conditions three to four replications have been found to be required to obtain reliable statistical estimates and provide enough degrees of freedom for this type experiment and the analysis of variance which would be used. A total of 12 to 16 experimental units (pens of cattle) would be required to execute the research.

> The nature of the problem under study suggests the statistical precision required in the experimental results.

An alternative to analysis of variance with similar results, is regression analysis. If this technique is envisioned for the feed additive experiment, the same four treatments could be used and if the interval between treatments was not uniform, they could be adjusted to make them equal (a desirable though not necessary attribute of a design for regression analysis). In analysis of variance, the observations from only two treatments are compared at any one time. In regression analysis the observations from all the treatments are considered simultaneously. For this reason roughly the same amount of statistical precision can be obtained with two replications of the four treatments for regression analysis as with four replications of the four treatments for analysis of variance. Hence, regression analysis may require fewer experimental resources than analysis of variance. Similar relationships hold for other types of analysis which may be required in the project and should be considered when choosing the design to be used.

The statistical precision required in experimental results is associated closely with the nature of the problem for which the research is being undertaken. Lower levels of confidence are acceptable in research dealing with animal health than with human health, for example, and lower confidence levels may be satisfactory in some social research than in some biological research. But regardless of the nature of the research, an increase in the precision required is almost always associated with the need for more experimental resources. For any given research objective or experimental design, more replications and/or more treatments will usually result in more precise statistical estimates than would fewer treatments or replications.

In addition to increasing the number of treatments or replications, precision can also be increased by closer supervision and better control during the experimental process. Spotty application of fertilizer by hand broadcasting on grass plots can result in large experimental errors as can carelessness and lack of thoroughness during harvest. More time, workers with more skill, or specialized machines or equipment can reduce this source of experimental error, but all will add to the requirements for experimental resources. From a practical point of view, the least expensive means of reducing experimental error from this source is closer supervision by the researcher during all phases of the experimental process. Time spent at the experimental site by the researcher can be highly productive, particularly when it is possible to prevent the complete loss of data through carelessness.

The use of prior information can reduce the need for resources by reducing the range and number of treatments otherwise required, by taking advantage of known estimates of variance to minimize the number of replications, and perhaps even by providing all the data needed making further experimentation unnecessary. In the following section we will discuss in more detail the use and utility of secondary experimental data, which is one source of prior information available to the researcher.

The time limit within which a decision must be made is also an important factor in determining the size and complexity of an experiment. If time is not a critical factor, an exploratory experiment of relatively simple design can be conducted and this can be followed by an experiment which is more exact in its focus. The range over which the experiment is conducted can be successively narrowed until the required accuracy is achieved. None of the experiments in this sequence need be large nor complex. But if timing is critical and decisions must be made rapidly, it may be necessary to increase the size and complexity of the experiment in order to include several links of the chain all at the same time. That is, to conserve time (in this case a very limited resource) other available resources may need to be substituted in order to achieve a useful research product within the time period allowed for making a decision.

Secondary Experimental Data

When experimentation is undertaken in the execution of the research project, the experimental design can and should be tailored to the needs of the project. The data obtained will then be the best available and will be well suited to the needs of the researcher. But it is not always necessary to conduct an experiment to have suitable or adaptable experimental data for analysis, because in most places where applied research is being conducted, at least some prior experimental data are available. These data, though generated for other research purposes, frequently provide an insight into the nature of the relationships to be studied in the current project, and may provide sufficient data so that additional experimentation need not be undertaken.

In situations where time is a limiting factor, the use of secondary experimental data may be advantageous or even essential, but data generated by another person or from one or more other experiments must be used with caution to assure that they are relevant and comparable. Some adjustment or selection may well be necessary to adapt the data to the current study.

Several means can be used to adapt, select, or adjust data of this nature. One means is to select only the relevant portions of the data and exclude those parts not related to the current analysis. An example would be to select data in a fertilizer experiment from those plots where potash was at a constant level, eliminate plots where trace elements were included, and make the analysis for the portion of the data in which only nitrogen and phosphorus were variable.

A second method of using secondary data is to analyze the results of several experiments and search for consistencies which may indicate relationships not otherwise evident. In the Cauca Valley, Colombia, production functions for a series of regional fertilizer trials in corn resulted, individually, in very poor statistical estimates. For any one trial, little confidence could be placed in the conclusions. But after dividing the trials into soil types, it was noted that there was a great deal of consistency in the functions within a soil type. The form of the curves was similar with each reaching a maximum within a short-range, and with calculated optimum applications at about the same levels. Hence, by using information from all the curves, together, general recommendations could be made even though the individual analyses yielded little information.

A third useful method of analysis is to consider the possibility of combinations of data from two or more different experiments which were conducted under similar conditions. This method was used to estimate the optimum stocking rate for fattening steers on grass from three different experiments [Gallo and Hildebrand]. Each experiment was conducted to determine the effect of hormones, and all were conducted on similar grasses, under similar conditions and with comparable cattle. Because the rate of stocking was different in each case, it was possible to use a mathematical response relationship from which the optimum stocking rate could be calculated.

> Experimental data generated by another person from other experiments must be used with caution to assure that they are relevant and comparable.

One additional point with respect to secondary data should be mentioned. When one has spent time attempting to use secondary experimental data, the productivity of any efforts made by the first researcher to preserve the data for other users is appreciated. Nothing is more frustrating than to discover a description of an experiment that should have provided usable information and then to find the data in such poor shape that they are impossible to use. Simple notations such as units of measurement are often not even included. The thoughtful researcher will leave a clear record of the data so that users who follow will have no trouble interpreting them.

Multipurpose, Multidisciplinary, and Multilocation Experimentation

As a practical matter, a great deal of experimentation is carried out with an orientation that is only partly research centered. An important example in agriculture is the "demonstration trial" usually conducted by, or in cooperation with, the extension service. One of the purposes of this type of research is to demonstrate under real conditions (i.e., under conditions which will be applied by the client toward whom it is focused), the results of research which was conducted under controlled conditions. Rather poor experimental control is to be expected, so accordingly, the experimental design is nearly always quite simple. In many cases, as few as two treatments, with and without a particular input or package of inputs, are included. At times, a complete experiment is conducted at one location; in other trials, different locations are considered to be different replications of the same experiment with all treatments being repeated at each; and in some cases only one or a few treatments are conducted at each location. Another useful concept is to consider each location an environment and carry out multi-locational research with the idea of predicting the response of the different treatments in varying environments.

It should be obvious that as more and more locations are included in a demonstration trial, exposure to the clients is increased, but experimental control is decreased. Also, with more treatments, more information is possible, but also, supervision becomes more difficult. Hence, the persons responsible for the trial must determine, based on the orientation of the project and the available resources, what the size and scope should be.

Farms as Replications

Presenting an example of a fairly successful demonstration trial may be the best means of discussing some of the more important aspects to be considered when initiating a project of this nature. This project was conducted in southwestern Colombia in an area not far from the city of Cali.[4] The area comprised that of a large number of descendants of former slaves who had divided their holdings into an average of about 2 hectares per family. The farms were nearly all in old stands of cocoa, coffee, and plantain which were in such poor condition that monthly income per farm barely reached fifteen dollars. General nutritional deficiency was evident as was the nearly complete absence of any source of animal protein. Physical conditions for any type of livestock were

---

[4]Peter Hildebrand, while working in 1970 with the Colombian Agricultural Institute (ICA) at Palmyra, Colombia, was one of the investigators with the Khaki-Campbell duck project.

very bad, and, of course, little or no money was available for supplemental feed. Conditions were quite homogeneous from farm to farm.

> Some research problems, settings and resource situations require multi-disciplinary, multi-purpose and multi-locational experimentation where farms serve as replications and/or environments within an on-farm research orientation.

As part of a development program for this area, it was thought that Khaki-Campbell ducks (an egg laying breed) could prove to be a potential source of animal protein to supplement the diets of these poor rural families. In designing the project, several alternatives were possible. The simplest of those considered reasonable was to select a small number of families (those most apt to be conscientious in maintaining the necessary records), to give each about 10 ducks and provide them with a recommended concentrate ration. The results could probably be established with a fairly high level of confidence and would show whether or not the ducks could survive under the conditions of the area. It would also be possible to determine the cost of the eggs to the families and whether or not they would eat the eggs.

A second approach considered was to use more families and experiment with several rations for the ducks. At first, three rations were considered adequate, with three families provided with each ration. In this manner, it was hoped to be able to determine not only if the ducks would survive and lay under the severe conditions prevalent in the project area, but also what the lowest cost ration would be. In other words, there would be three different rations to choose from. Although more information could be obtained, more families would be needed and more supervision required to assure that the rations were correctly administered. The extension personnel who were cooperating in the project felt they could find nine families and had sufficient resources to do the majority of the supervision. Over 100 young female ducks were going to be available, so the extension people began to locate the cooperating families.

As it turned out, 16 families were eager to cooperate in the project, and the extension people felt this number should not be too great a supervisory burden. The use of this many families would reduce the number of ducks available per family to seven. It was decided that this would be satisfactory, but that it would be more efficient in the use of the families to include four rations and have four replications of each. With four rations it was hoped that there would be sufficient information to estimate a production function for eggs and with this information be able to determine the lowest cost ration even if it were not one included in the experiment.

The final design included one group with a complete concentrate ration, one with 2/3 of this amount, one with 1/3 of a ration, and a control group with no concentrate. Except for those ducks receiving a full ration, all could receive whatever scraps were available and be allowed to graze (or be fed chopped grass). Material for constructing adequate housing for the ducks was available locally at no cost and all families were required to be ready to receive the ducks by a fixed date.

A one day short course was held with all families participating sufficiently far enough in advance of the delivery date to allow time for construction of the housing. The care of the ducks was discussed and also the design of the experiment and its purpose. The families were all given very simple record forms and instructed in their use (to be able to use the forms it was not necessary to be able to read or write or even count). Cups were provided to measure the exact amount of feed for each duck for each ration so that in the event of losses, the adjustment of the ration would be simple. At the close of the course a meal including duck eggs was served to assure the families that the eggs were good to eat since there was some local bias against duck eggs.

With the encouragement of the extension agent, all participating families were ready on the date that the twelve week old ducks were delivered. Initially, extension field persons and the researchers checked frequently with the families to ascertain if everything was being done properly. The records were collected every two weeks during the whole project to assure they were kept current.

To be sure, there were rather large differences in the results for each ration. One family, for example, never was rewarded with even one egg. Although they were a bit embarrassed and the neighbors were sure they were not taking good care of their ducks, they took it good naturedly because they were assured that this was not unexpected and was a necessary part of the experiment. Nevertheless, by having four replications adequate information was obtained. It was possible to determine the best (maximum profit) level of concentrate to use if the eggs were going to be sold, as well as the ration which resulted in the lowest cost for the eggs if they were to be used to supplement the family diet. The second use, of course, was the primary purpose of the project.

Upon completing the analysis of the results, a field day was held for the families at which time the results were presented. After the general results were shown, those of each family were discussed in an attempt to determine why some had better results (hence lower costs) than others with the same ration. Following this discussion, all the families participated in the selection of the final recommendations which were published.

In summary, this project was tailored to fit within the resources available and yet to provide the greatest exposure in the project area. Although experimental control was relatively low, sufficient treatments and replications were included to maintain adequate statistical reliability in part due to the presence of a high degree of supervision by the researchers.

Farms as Environments

As mentioned previously, farms vary widely. For technology evaluation results to be useful to large numbers of farmers, there must be a way to predict responses for different kinds of fields or farms. An efficient means to do this is to conduct research on a number of farms. One possibility is to use a group of similar farms as replications so that results will apply to that kind of farm. This would require several groups of farms if results were to apply to more than one kind of farm or type of field. Another alternative is to use a wide array of farms or fields to differentiate responses by farm or field characteristics. An efficient procedure of the latter approach is with Modified Stability Analysis [Hildebrand, 1984]. This procedure utilizes an environmental index calculated for each location. The index is equal to the average yield of all treatments at that location. This index reflects whether the environment on that farm or field is "good" or "poor" for the particular crop or product being evaluated and for the treatments included in the experiment or trial. It is efficient because many fewer locations are required than would be the case if all factors influencing environment were to be included as independent variables in multivariate regression. Interaction between treatments and environments can be evaluated by this procedure.

On-farm research in the Amazon rain forest of Brazil [Singh, 1990] will serve as an example. Two small farming communities that practice slash and burn agriculture in the municipality of Rio Preto da Eva in the state of Amazonas, were selected for this project. Following a study of secondary information and a Sondeo, three treatments to help solve the problem of low soil fertility were selected from previous on-station research. They were compared with farmers' practices for maize (*Zea mays* L.) and cowpea (*Vigna unguiculata*). Partial results from the cowpea trials are shown in Figures 6, 7 and 8.

The four treatments were the individual farmers' practices (FP), chicken manure (CM), processed city waste from Manaus (PCW), and triple super phosphate (TSP). The trial was planted in a randomized complete block design with two replications where possible (some locations had only one replication because of limited area allocated by the farmers for the trial). Soils upon which the trials were planted were on land taken out of primary forest and farmed 1, 2 or 3 years (PF_1, PF_2, PF_3); or secondary forest and farmed 1, 2 or 3 years (SF_1, SF_2, SF_3); and land farmers classified as wasteland (WL). For cowpeas, 13 locations were included in the analysis.

> The environmental index reflects whether the environment on a farm or field is "good" or "poor" for the particular crop and treatments under evaluation.

Figure 8.1 shows the relationship between the environmental index (EI) and the four treatments using mg ha^{-1} as the evaluation criterion. If a farmer is interested in quantity per ha, then the best treatment for the better environments (EI $>$ 1.4) is TSP. In the poorer environments, CM results in the highest yield. The relationship between soil types and environmental index is shown below the graph in Figure 8.1. Land taken out of primary forest and in the first or second year of production and land from secondary forest and in the first year of production is associated with the higher environmental indices. All other land types are associated with the poorer environments. This information allows the researcher to make specific recommendations with respect to land type.

Figure 8.2 shows the relationship between the environmental index and the four treatments using kg per dollar of cash (kg \$$^{-1}$) required for the treatment as a criterion. Notice that this gives very different results than when Mg ha^{-1} is used. Farmers who have very little cash and must

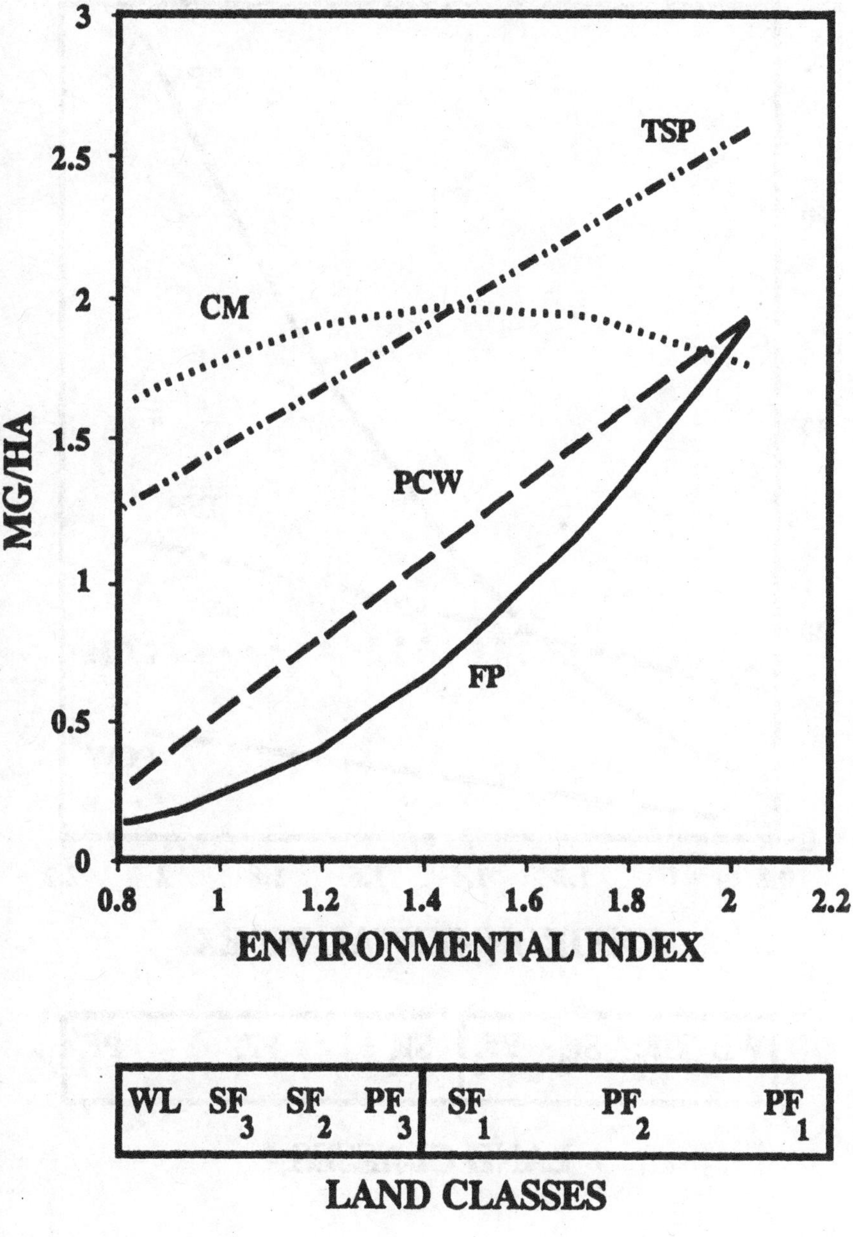

FIGURE 8.1 Cowpea response of four treatments to environment (with mg/ha as the evaluation criterion)

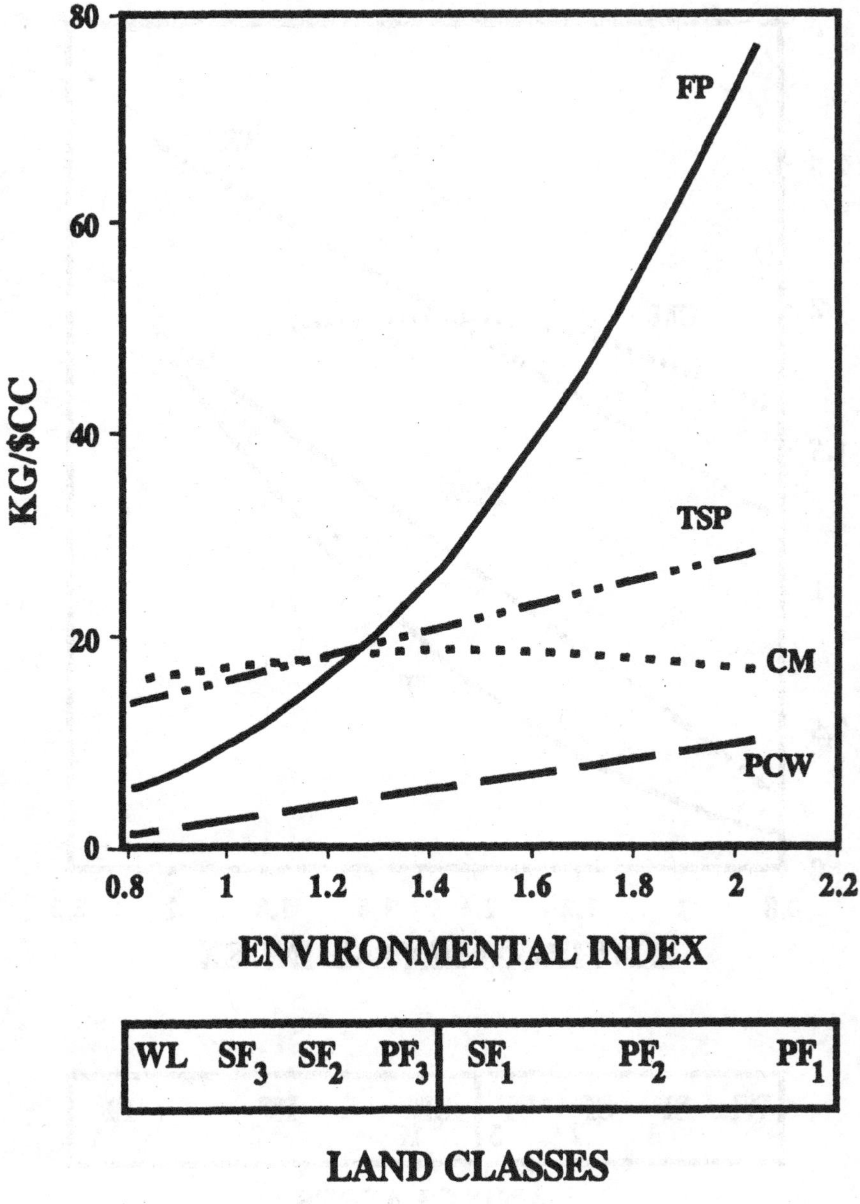

FIGURE 8.2 Cowpea response of four treatments to environment (with kg/dollar of cash as criterion)

maximize its productivity will not use a soil amendment if they have land taken out of primary forest and in first or second year of production or from secondary forest and in first year of production. If they are forced by population or other pressures to use other land classes, then they may use amendments if available. Either CM or TSP would appear to be a reasonable recommendation.

Additional information upon which to base a recommendation can also be obtained from these on-farm data. Figure 8.3 shows the distribution of confidence intervals for kg $\$^{-1}$ for the 6 locations in the poorer environments. This evidence indicates that CM is a superior treatment to TSP.

In summary, this type of on-farm research provides a much greater amount of specific information for the clients than does the more "precise" research carried out at only one location (an experiment station).

Multidisciplinary Experimentation

> Multidisciplinary experimentation is a means of efficiently using scarce research resources.

Multidisciplinary experimentation is a means of conserving scarce research resources. When researchers from two or more disciplines cooperate in a project it is frequently possible to obtain answers for each researcher with little change in the basic experimental design.[5] Without collaboration, a researcher in one discipline often minimizes the effects of those factors commonly included by researchers in other disciplines, so that in the absence of cooperation, the product is of low value to other researchers. An example occurs in beef or dairy projects where the researchers raise their own maize for silage without the cooperation of the maize researchers who may be at the same experiment station. The animal researchers may be able to document accurately the effect of the silage on the animals but be unable to estimate animal production per hectare of maize because they had little interest in the maize

---

[5]For a brief and excellent discussion of multidisciplinary cooperation in extending experimental results to practice, see McPherson.

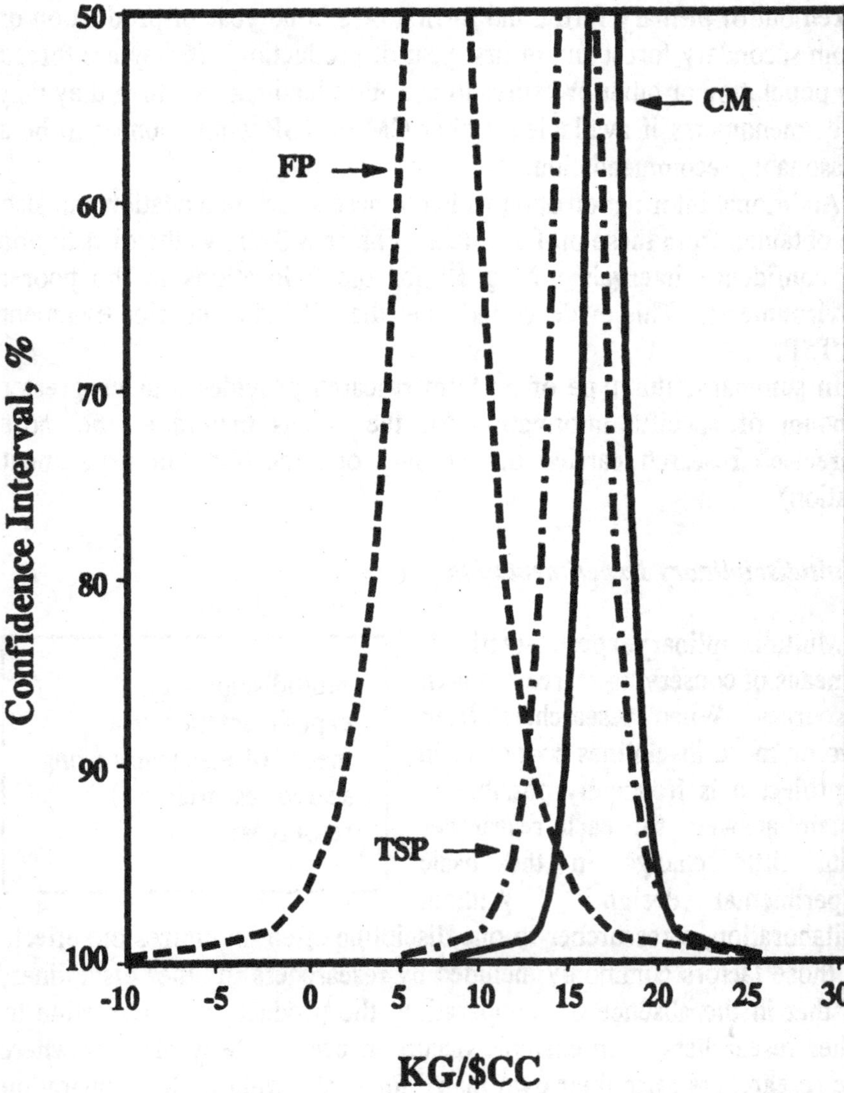

FIGURE 8.3 Stability of three treatments in cowpea for poor environments

production. Even more difficulties will arise if it is later desired to make an economic evaluation of feeding silage to animals [Escobar].

Of course, it is not always possible to obtain answers for more than one discipline without increasing the size and the complexity of the experiment beyond manageability and available resources. But it is precisely in those cases when resources are most scarce that it is important to consider the advantages of multidisciplinary research. Most experiments are costly both in terms of time and of other resources so it is obvious that if the results can serve two or three researchers, the additional effort required in planning the experiment can be extremely valuable [Kiehl].

Summary

Experimentation and experimental design usually are associated with objectivity, precision, and scientific purity -- concepts that imply rigidity and inflexibility in thought and procedure. In basic research this is mostly necessary, but in applied research, considerations other than pure scientific objectivity can become more important in determining the type of design to be used for any given experiment. Resource limitations will often force a reduction in the accuracy obtainable, and factors such as demonstration uses will modify the nature of the design ultimately selected. It is important that applied researchers maintain a flexible attitude with respect to experimentation and experimental design in order to increase their effectiveness and make them more efficient in their work.

9

Behavioral Research Based in Interviews

This chapter will first focus on the selection of informants or respondents and the design and implementation of information gathering techniques and then discuss verification and preparation of both primary and secondary information for analysis. Some emphasis is given to structuring interviews and for case studies. Time and resource restrictions along with the needs imposed by the hypotheses and objectives will dictate whether the measurement process will be a large survey, a group process or a case study [Sudman].

The non-experimental information gathering techniques discussed will primarily cover the conventional interview from a questionnaire. Questionnaire use is considered to be a "closed-ended" approach in that the structured nature of the questionnaire, even when qualitative or subjective questions are asked, limits the process to a very specific set of concerns. Research resource limitations often dictate how "closed" the questionnaire will be. More "open-ended" approaches including case studies, focus group interviews, multidisciplinary team interviews, and cooperative inquiry have been touched on previously.

> In non-experimental research, the researcher usually cannot deter-mine design and level of the variables nor directly measure the phenomena, but controls only the measurement technique.

The distinction between the terms "respondent" and "informant" is important. When closed-ended questions are used in a survey, the persons who are answering can only *respond*, "yes," "no," "5," "last year," etc. Thus, questionnaires structured with closed-ended questions rely on "respondents." Because each respondent answers the same question, results can be analyzed by precise statistical procedures. However, if the list of questions is not complete, information will be lacking.

When open-ended questions are used in a questionnaire, each person has an opportunity to augment or modify the information given. More insight can be gained but the information received from the "informants" cannot be analyzed as precisely. Finally, in some kinds of surveys the researchers are searching much more for insights than for statistical analysis. They are interested in "informants" and may not use questionnaires at all.

Conventional Interviews from Questionnaires

Selection of Respondents

The selection of the respondents, or sample design, is an important part of the process of non-experimental data collection. Because the researchers have no control over the distribution of the factors they are studying (these exist as a result of nature and social organization), the proper selection of the respondents is necessary to assure that the information is being obtained from an appropriate population. Once respondents have been selected and the interview process is completed, a faulty or ill-designed sampling procedure may generate data that do not describe the target population. To resolve this difficulty more time and resources, if available, must be expended than would have been necessary with a more carefully designed sample.

Two commonly employed sampling approaches are random and stratified samples [Brauer and Hunter; Cochran, 1963b; Hansen, Hurtwitz and Madow; Kmenta]. For information of a broad census nature, a completely random sample gives each person in the total population (all farmers, high school students, married women, etc.) an equal chance of being chosen. In this manner, results can be associated with the characteristics of that population. Stratified samples, where particular sub-groups are chosen, are used to assure that each sub-group is equally represented, or to assure uniformity or representation over specific ranges of certain group traits (different farm size or different income groups for example). Such a stratified sample can approximate experimental measurement but with less precision than is normally associated with experimental research. A stratified sample is often more efficient than a random sample in use of scarce research resources because the sample size required may be smaller reducing interview, data storage, retrieval and processing expenses. However, the variables selected for stratification are very important and must be chosen with care and based on knowledge and experience.

In drawing a sample, one hopes to have available a reasonably good listing of the general population including some descriptive characteristics to aid in stratification. To help determine the sample size, measures of variance within the population may then be calculated and used along with resource limitations and levels of confidence desired by the researcher and the client.

Many times in applied research, these general characteristics of the population are unknown, making sample selection more difficult. One researcher studying isolated farms in a rain forest was forced to utilize somewhat unorthodox but otherwise relatively functional techniques [Feaster]. In one of the three zones of the region to be studied, maps were available showing farm locations. A simple random sample of farm households was easily obtained from these maps. In another zone "trail maps" were available but the farms were not shown. The approximate total number of farms was known and the trails and farms were, in general, uniformly distributed throughout the region. A complete enumeration to develop the sample frame would have been difficult and would have required more time than was available. The researcher chose to send out interviewers by mule with instructions to obtain an interview at every fifth farm. The sampling problem was thus handled as the interviews were taken. In the third zone, maps were extremely poor and little was available to guide the sampling process. To assure that the entire zone would be represented, the interviewers were instructed to ride their mules for one hour and then take an interview at the nearest farm, again proceeding along the trail. As one might expect, errors did result, but adjustments in the analysis were based upon general observations made by the researcher while in the zone.

Had logistical problems not been so extreme, the above researcher could have used a sequential sampling technique. That is, based upon a predetermined level of confidence, he could have measured the variance in selected population characteristics by using the interview data taken daily while in the field. From this calculation he could then determine when the sample had reached the size needed by reducing the variance to meet the desired confidence level. When that level had been achieved he could have moved to another zone.

> Flexible, applied, non-experimental researchers place emphasis on the research problem and view their sampling techniques as tools and not ends.

A further need for flexibility in the sampling process to allow for obtaining needed information is illustrated by a research project in East Uttar Pradesh of India [Jhunjhunwala]. Initial interview attempts with Indian farmers concerning their land holdings were met with considerable suspicion and apprehension, possibly because of speculation about land ceiling legislation. Interviews conducted in villages as planned, encountered farmers unwilling to provide the desired information. Yet when the farmers were selected at random while traveling along the roads in the district, they responded readily to questions about village and other matters related to land ownership. Since personal identity and specific location of the respondents were not imperative, the sampling approach was changed substantially but to the benefit of the entire research effort.

In summary, flexible researchers who place major emphasis on the research problem and view their sampling techniques as tools and not ends in themselves will effectively perform applied non-experimental research under even much less than optimum conditions.

Designing the Questionnaire

At the risk of seeming elementary, it is advisable to emphasize that questionnaire development must be closely tied to the problem, hypotheses, objectives, analytical techniques, and available resources for the proposed research. Just as the research proposal or plan must reflect a relevant problem, and the hypotheses and objectives must pertain directly thereto, the respondents to be interviewed and the questions to be asked must be relevant to the research proposal. Beyond relevancy, a questionnaire must include only those questions that are highest in priority for analysis and testing. A multitude of questions could be "relevant" or "interesting" but only a limited number can be used fully in the project because of research resource limitations. Interview time, for example, represents a costly resource allocation not only for the research project, but also for the respondent. The researcher must always remember that excessive and/or irrelevant questions try the respondent's patience, reduce interviewer-respondent rapport, threaten credibility of the responses, and waste scarce resources of the research project.

Difficulties in Interpretation and Communication. The method of obtaining information in a questionnaire is the spoken word, so language problems must be given special consideration. Obvious problems arise when questions are translated from one language to another, but even within the same language, word and expression usage vary with different cultural, social, educational and economic classes. Differences of interpretation may exist between the researcher and the interviewer as well as between the interviewer and the respondent. There also will be differences between respondents in the interpretation of some questions. The magnitude of this difference will increase as the heterogeneity of the sample increases.

The complete elimination of language problems is nearly impossible, but careful consideration of cross-cultural communication difficulties will minimize problems of interpretation in the questionnaire. Two major cross-cultural communication problems in questionnaire formulation are differences in terminology and cultural differences in beliefs and values.

The names by which things are known or the terms which are used to describe something vary widely even within a language and culture group [Hall]. These differences in terminology must be considered in the construction of the questions. Generally, the lower the general level of education and the greater the degree of isolation, the more important it is to take local usage into account. When groups are isolated, indigenous objects and phenomena frequently have names which are known only within the groups. Objects or phenomena which are not indigenous will probably have similar names, but if they are unknown within the group, they may be difficult to explain. Terms used by professionals may not be understood at all by the respondent, yet if the interviewers are allowed freedom in explaining these unknown terms, an interpretational bias may arise. In some cases, professionals may be able to communicate with each other and with the interviewers with one question, but several questions may be necessary to obtain the information from the respondents of a survey.

A common example of the problem with varied usage is with land measures in Colombia. Four measures, the hectare, the plaza, the cuadra and the fanegada, are used in different parts of the country. When a survey is conducted, one must construct the questionnaire to account for this variation or risk interpretational errors in measuring farm size, crop yields, and other factors associated with land area.

For effective research resource use, questionnaire design and implementation must deal with relevancy, language, cross-cultural credibility, measurement and other challenges.

Another example is a case that arose in Colombia with respect to a question concerning tapeworms in swine.[1] In each of three departments (states), a different local term had to be used in the question to avoid erroneous conclusions which could have been drawn when determining the respondents' knowledge of the incidence of this parasite in the survey area.

Unknown or misunderstood differences in beliefs and values are the other important cause of interpretational problems. An example can be drawn from interviews with housewives in a Colombian village where the drinking water was badly contaminated [Rogers, pp. 7-12]. Many of these housewives realized the importance of boiling their drinking water and understood the technique involved but still did not do it. Informal interviews determined how many housewives did boil water, how many realized they should, and how many understood the technique, but the survey failed to determine why the majority did not actually boil water. The difficulty developed because appropriate questions were not asked due to inadequate cultural knowledge. The professionals assumed that boiling water was "good," and based upon survey results and their own normative considerations, they determined that housewives "should" have been boiling water. Only after further questioning did the researchers discover that the villagers believed that water was boiled only for the very ill. Those who drank boiled water were considered to be ill, so to avoid this stigma, many who otherwise might have boiled their water failed to do so.

Designing for Data Retrieval. A questionnaire must lend itself to efficient and useful data collection, processing and analysis. The allocation of time and resources between field collection and data retrieval and verification from completed questionnaires needs careful

---

[1]The question is from a questionnaire which provided information for a Masters thesis [Gallo].

consideration when designing a questionnaire. Long and detailed qualitative responses to open-ended questions may be more informative than quick semi-quantitative responses involving selection of alternatives, but the time required to correctly obtain, retrieve and analyze one long response may be the same as for twenty short selective response questions. One must determine which of these two techniques (or combinations of both) most efficiently provide information while minimizing interpretation and response errors.

Both collection and verification are facilitated by logic and consistency and by organizing the questions to lead the respondents with ease through the questionnaire in a manner that stimulates spontaneity in their responses. If one question asks about a farmer's marketing problems for rice, and is followed by the age of each member of his or her family, and the next returns to the price of rice, the interview becomes burdensome and less effective. One exception to this rule is important. If one desires to cross-check or verify a particular response because of interpretational problems or a reluctance to respond accurately, it may be desirable to ask the same question twice, each time in a different manner and at different points in the interview.

Dividing the questions into major groups will help reduce respondent fatigue while maintaining interest. Groups of questions that involve sensitive issues, such as income or profit that may be subject to taxes, might follow several relatively less sensitive groups of questions in order to establish rapport and confidence. Sensitive issues covered prematurely in a questionnaire may destroy the entire interview and thus should be asked toward the close of the interview.

Coding, or preparing the data for computer analysis, is usually necessary. Precoding can eliminate the need to transfer the data to code sheets by permitting the data to be transferred directly to computer memory. This method requires more careful editing of the questionnaire but usually reduces retrieval costs and the chance of human error by removing one transfer of data. To be even more efficient, however, researchers must be certain beforehand of how they want to organize and process the data. The editing and checking process, on the other hand, helps assure that all necessary data will evolve from the questionnaire. Precoding is less advisable for highly qualitative responses or when the researcher has no indication of the range of responses expected for each question.

Pretesting the Questionnaire

Pretesting the questionnaire or checking to see if it will obtain the information sought, must be accomplished under actual field conditions before beginning the general survey. No amount of intellectual exercise in the office can substitute for properly testing a questionnaire among the respondents in the area to be surveyed. This phase will determine the weaknesses in the questionnaire, establish whether the information sought can be obtained in a useful form, and also may provide additional information which can help improve the relevancy of the questionnaire. It is often useful to analyze the data obtained in the pretest to assure that the form in which the data are obtained make them amenable to the desired analyses. Preliminary estimates of variance can also be obtained to help verify sample size.

Whenever possible, the interviewers who will be conducting the survey and the director of the research project should be involved actively in the pretest to assess problems with the questionnaire and with respondents. Prior to pretesting, interviewers should become familiar with the research project and the questionnaire. Each interviewer should be briefed carefully and sent to several of the different areas which have been selected to assure experience with a variety of respondent types. When each interviewer returns, a debriefing with the researcher should be performed immediately. This debriefing should include a review of the completed questionnaires to identify points of misunderstanding, unnecessary repetition, time difficulties and so on. General reactions by both the respondents and the interviewers concerning the subject matter and length of the questionnaire should be discussed.

For some interviews care should be taken in selecting the areas where the pretest will be performed to avoid the danger of contaminating a major target area by pretesting the questionnaire within its bounds. Especially in close-knit rural areas, advance knowledge of attitudinal responses in the final survey could be affected by word of mouth discussion among potential respondents and those participating in the pretest.

Size of Pretest. The required number of pretest questionnaires depends upon the research problem, the homogeneity of the survey population, the data collection and analytical techniques to be employed, the total number and complexity of the questions to be asked, and the research resources available for the project. The range may be from five to one-hundred or more questionnaires. If the population is very heterogeneous with respect to information required, more interviews will be necessary for an adequate pretest than for a questionnaire to be administered to a more homogeneous population.

If the pretest is also to provide desired information about variance to determine the population homogeneity for sampling purposes, more pretest interviews may be needed than when only a test of the questionnaire is desired. For potentially large studies where general population parameters are unavailable, a rather extensive pretest can aid in designing the sample. The marginal costs associated with these additional pretest questionnaires must be measured against costs associated with sampling errors due to having too few respondents. Costs resulting from obtaining excessive questionnaires in the main interview effort may be greater than the costs for a few additional questionnaires in the pretest.

Information Checking. How a pretest improved a particular question in a survey of settlers in a colonization project in Caqueta, Colombia is illustrated as follows [Feaster]:

Original question:

Where did you live before you moved here?

Improved questions:

In which Department were you born?

Where did you live before moving to Caqueta?

Where did you live before moving to this farm?

The researcher wanted to know from which departments the settlers had migrated to Caqueta, but the original question only revealed information about the respondent's last move which may have been within Caqueta and not from another department. Even though questions were added, the results were easier to interpret. Based on this researcher's information needs, the first question had no value so he was forced to choose between no question and three questions.

A pretest will assist the researcher to develop an efficient but complete questionnaire. A poorly pretested set of questions designed to obtain information about farm management practices and production costs for potato producers in Colombia illustrates this point.[2] The respondents

---

[2]From a portion of a producer questionnaire used in Colombia and summarized in Andrew, 1969.

were asked the number of hectares devoted to potato production for the most recent year at three different points in the questionnaire. At the beginning, producers were asked to give total farm size and land use by crops, pasture, and other classifications. Later in the interview they reported on potato plantings for each semester. This information often did not agree with the first question because of double counting. It was difficult to verify the response because plantings in the first semester were usually much larger than in the second, so the extent of the double counting was impossible to determine. And finally, when queried about hectares harvested, the area given was usually slightly different from that previously used. Could it be assumed that those hectares not harvested were lost to blight or drought? Perhaps, but it would be better to be certain that the difference was not in part due to either intentional or unintentional reporting error. Not only was the duplication unnecessary, but the questions were ambiguous. A better pretest could have minimized this discrepancy by guiding the preparation of one comprehensive and easily understood set of questions related to hectares of potatoes planted and harvested for the past year.

After completing the pretest, responses should be tabulated and critically reviewed to determine whether the questions were understood clearly and whether the information which resulted would help to resolve the problem. This step frequently uncovers gaps in information, or responses which are not in the best form for the analyses to be undertaken. While the pretest may identify needed modifications in the questionnaire, it also provides an opportunity to determine whether the responses are in a form to test the hypotheses. The researcher should carefully consider what is being collected in the pretest and how it can be used in the hypothesis testing and analysis phases of the research. Pretest data can be subjected to a "practice run" through the analyses. This can aid greatly in refining analyses and it may also provide some ideas as to the format for the final publication. If these benefits are gained from analyzing pretest data then the saving of time at a later stage in the research process will more than compensate for any early time loss.

Often a pretest indicates that either too little or too much information is being collected. At times, the pretest may show that a certain type of information will not be available, and therefore, some modification will need to be made in analyses, hypotheses, or objectives. More commonly, the questionnaire may be gathering more information than is necessary which suggests that costs of the excess should be carefully compared to potential beneficial uses of the information.

Only when pretests have shown that interpretational problems are minimal and that the information can be obtained in the form required, should the questionnaire be finalized and the complete survey undertaken.

Time Difficulties. Because timing is so important, the pretest should be used to help specify interview time problems; time of year, month, week, and day must be carefully considered to fit the respondents' work schedule.[3] If, for example, one is proposing to interview truckers at check stations or consumers at a retail market the time element may be more critical than in interviews with producers. Retailers usually prefer granting interviews when the fewest customers are in the store. For small retailers, a daytime interview can take three or four times as long as an uninterrupted night time interview; but if an appointment cannot be obtained for nonworking hours, as is often true, the more lengthy daytime interview may be necessary. Agricultural producers probably will find planting, weeding and harvest times to be inconvenient.

Numerous examples can be cited but few guidelines can be specified for timing interviews because each survey and often each interview represent different situations. In any case, should the respondent feel the questionnaire is completely irrelevant and a waste of time, the responses will lack credibility even when the interviewer is fortunate enough to complete the interview.

Selecting and Training Interviewers

Even a well-designed and pretested questionnaire will not yield good results if administered by an interviewer who is poorly trained or who

---

[3]Mail questionnaires help resolve time problems but often limit response quality and, due to low response rates, require large samples. A bibliography concerning mail-questionnaire research is found in Buse. Work in the 1980s to improve the effectiveness of mail and telephone surveys has been quite successful in raising response rates. Due to survey interview costs associated with personal interviews this emphasis is improving primary data collection efficiency where the approach is suited to the particular problem under study [Dillman; Sudman].

has a disagreeable, non-cooperative and culturally or class biased attitude toward the respondent. The contact made with the respondent will not only influence the interview but also influence the respondent's evaluation of the organization for which the interviewer is working, and acceptance of action programs that might evolve from the research recommendations. Requisites of a good interviewer include: (1) an interest in and an understanding of the research project, (2) an interest in people and the ability to communicate that interest and sincerity to a respondent, (3) a willingness and ability to follow instructions and definitions without regard for personal beliefs or convictions, and (4) in general, a good public relations attitude.

> No amount of intellectual exercise in the office can substitute for properly testing a questionnaire and field training interviewers.

Training is necessary to familiarize experienced as well as inexperienced interviewers with the problem to be studied, the objectives of the research project and the organization sponsoring the work, the questionnaires, the respondent selection procedures and the interviewing techniques. An interviewer's manner of asking questions must be objective and the interviewer must be neutral and honest in recording responses. The most important goal of the survey is to get an unbiased opinion or bit of data for each question. Whatever the interviewer thinks of the respondent, his or her opinions should not influence interviewer objectivity.

Interviewers should be instructed carefully on methods for introducing the research project to the respondent. A most difficult question to answer is a respondent's inquiry about the legality of the study. Letters of introduction and any other documents from official and credible sources always should be in the possession of interviewers. To establish respondent confidence, the introduction used by the interviewer and the letter itself should explain that the interview is confidential, and that the information sought will be used to determine attitudes and characteristics of groups for comparative purposes, and not to illustrate characteristics of specific respondents.

Although the interview must be objective, interviewers must be given flexibility to obtain a useful response. They should record specific comments and not vague or meaningless generalities such as "because it is interesting" or "I like it because it is good." "Why is it interesting?." "Why is it good?" The "I don't know" response is one of the most difficult to manage. One is not sure whether the respondent really did not know, whether he or she did not understand the question, or whether the person did not want to respond for various reasons. Similarly a simple guess or an estimate calculated to suit what the respondent feels the interviewer wants should be avoided. Respondents, however, should not be encouraged to give an "I don't know" response when they most likely can give a meaningful and honest reply.

Personal, Telephone and Mail Interviews

Cost considerations and the particular type of information desired influence the type of information gathering technique employed. At one time neither telephone nor mail interview techniques were capable of delivering qualitative and quantitative responses acceptable to researchers and their clients. Improved practices and techniques, when properly employed, are now available that tend to raise response rates and the quality of response. In countries such as the United States people are conditioned to various opinion poll and interview activities through personal experience and exposure in the media. The product of these responses appears daily in many forms and tends to influence behavior of government and commercial entities. Skilled problem and opportunity identification, along with effective mail and telephone techniques often support varied causes in an objective and ethical manner. As people become more tuned to improved problem specification, alternative solutions and the process of information retrieval, along with sound techniques, they are more receptive to the mass mail and telephone interview processes. Advances in computer technology support these efforts. Dillman gives an excellent overview of mail and telephone interview experience and guidelines for successful information retrieval.

Verifying Primary Data

Verification of the data after they are collected is necessary before data processing and analysis begin. Researchers must be fully aware of the limits and potential uses of their data based upon the objectives of their research. A careful review of data with appropriate revisions at this stage of the research process will reduce the chance of costly errors.

Numerous and detailed verification activities are necessary so the researcher can be confident that the data best measure the phenomena needed to fulfill the research objectives, given the resource limitations placed upon the project. The verification and preparation process must consider techniques for handling missing observations, falsified data, inaccurate measures, and so on, as well as the necessary conversions where measuring units have differed. Computer processing is not necessary for all research, particularly when projects are small and calculators are readily available. Both time and monetary costs for each alternative should be considered along with the type of analysis desired and the degree of accuracy needed. Regardless of the processing facilities and techniques to be employed, well specified methods for verifying and coding data will improve efficiency in research resource use.

> Verification is necessary so that the researcher can be confident that the data best measure the phenomena needed to fulfill the research objectives.

During field data collection and while verifying and coding results, the researcher should accurately record the collection and verification techniques utilized. This record, commonly called a code book, is also oriented to describing data collection and verification techniques. It not only serves as a guide to present work, but also serves to indicate in detail what the data represent for use in further research efforts.

Summary

The success of non-experimental data collection rests upon the ability of the researcher to accurately and effectively select informants or respondents from the defined population and to communicate effectively with those selected. In designing questionnaire or other information retrieval strategies, various communication problems must be considered such as cultural, language and locational differences. Pretesting a questionnaire or an interview/exchange activity under actual field conditions can provide information about methodological effectiveness of the data gathering tool and provide information on the population which can help in establishing respondent selection criteria and sample size.

As in the case of experimentation, guidelines for optimum participant/informant/respondent selection procedures and interview design in non-experimental research can be stated, but in applied research it is important that the researcher remain flexible in his or her attitude. Scientific perfection can serve as a norm, but researchers must remember that for their clients, it is almost always better to have some information to help them make their decisions than to have no information except that the researcher is still designing a better questionnaire or trying to decide on the best means of choosing the respondents.

The range of activities classified within non-experimental work is extensive and not well defined. The emergence of various open-ended and qualitative methods for eliciting client needs can, through improved problem identification, help improve effectiveness and efficiency of other quantitative experimental and non-experimental information gathering methods. An applied researcher, much as a baker, must blend the ingredients in a fashion that will create palatable applied research results for the client.

research it is important for the researcher to remain flexible in his or her attitude. Scientific procedures can serve as a model, but researchers must remember that [illegible] systematic [illegible] to have more information on which to make their decisions, than to act on no information, except that the researcher is still designing a better questionnaire or trying to decide on the best strategy of choosing the respondents.

The range of activities [illegible] research [illegible] exhausts [illegible] The emergence of [illegible] and qualitative [illegible] problem identification [illegible] quantitative experiments and [illegible] experimental [illegible] techniques. [illegible] applied research [illegible] must [illegible] make available applied research results for the client.

10

Data Utilization and Information Synthesis -- What Does It All Mean?

The real skill of the applied researcher comes into play after the collection of the data has been completed. It is here that YOU as the primary research resource must draw from your full capability as discussed in Chapter 3. Your experience and imagination have a particularly high payoff in the analysis and the interpretation of the data and YOU make a difference between a useful project and one which ends up in a file drawer. It is in this process that you finally come down to the point of determining what the data entail; data do not "speak for themselves" but must be analyzed and interpreted. You, the researcher, must draw conclusions from the analysis and in the end make recommendations to the client that will help in resolving the problem which originated the project. This, of course, is the reason for undertaking applied research in the first place. No amount of planning, no elegant data collection procedures, and no sophisticated analyses are going to help researchers who are too timid when the moment of truth arrives to utilize all their information, draw meaningful conclusions and make appropriate recommendations to the client.[1]

> Finally when the moment of truth arrives, the client is expecting a useful product from your information gathering, retrieval, analysis and interpretation.

---

[1]For references particularly concerned with interpretation and communication of research results see Mann; Caunder; Holt, Pugh and Brant; Shaffer.

When this moment arrives, the client is expecting a useful product and the researcher is the most knowledgeable person available to the client. At the conclusion of a project, the researcher should know more about the subject being studied than anyone else with whom the client has contact. If this is not so, the client should have gone elsewhere for information. And if the client had not needed the information he or she would not have contacted the researcher nor invested the other research resources. Hence, researchers must assume that their knowledge is vital to their clients and that clients desire the fullest utilization of the resources which have been expended by the project. Also, because of the nature of applied research, the researcher is usually facing a deadline, so additional data collection is seldom possible. Conclusions and recommendations must be made on the basis of the data at hand because that is the best information that is or will be available within the allowable time and resource restrictions.

Rather than cover the myriad of analytical procedures available to the researcher, which are presented in detail in a variety of good sources, this chapter treats the more personal aspects involved in the interpretation of research results.[2] These are the aspects which might be called in part, the art of research, and which also might be called subjective analysis or, by the purist, personal viewpoints and judgements. We prefer to think of subjective analysis as flexibility in one's attitude toward the scientific procedure and the approach discussed in this book. This is the attitude that allows the researcher to look beyond the numbers which result from the analysis. It is this attitude that allows the researcher to completely utilize the data, drawing out all the information which might be of help to the client. Furthermore, this attitude encourages the good applied researcher to insist on a role in the interpretation of the statistical or other analyses which have been used (either by the researcher, by another person, or by a computer) rather than accept these impersonal results without question.[3]

Applied research is not useful to the client when the researchers report that they had to go back for more data and, as a result, have no conclusions, or that based on some specific level of confidence there is

---

[2]A sampling of basic texts covering analytical techniques includes: Ezekiel and Fox; Leontief; Peng; Brownlee; Theil.

[3]For a lengthy debate concerning the role of social science research in prediction and prescription see Christensen; Kelso.

no significant relationship among the variables. It is more useful to report that, although a relationship is not strong, and high confidence cannot be placed in the conclusion, there is a tendency toward a particular relationship and that because this response is logical, the relationship can be used carefully in resolving the problem. Or elsewhere, just as strong a statement could be made for there not being a relationship between certain variables. The client is depending on the researcher to draw a conclusion and make a recommendation. The client, then, armed with this best estimate of the researcher, ultimately makes the decisions to be taken to alleviate the problem.

> Now YOU are the most knowledgeable person available to your client and because the client must make a decision, your product must be usable and understandable.

Because the client must make a decision, it is also necessary that he or she understand the information which the researcher presents as the results of the research. Too often investigators write their reports as if they were communicating only with other professionals and ignore the needs of the client toward whom the presentation must be directed.

In this chapter, two factors associated with the utilization of data in applied research are discussed. First, an attitude of flexibility in the analysis of data prevents researchers from becoming constrained by tradition to the point that they are unable to understand what their data are trying to tell them. A lack of flexibility on the part of researchers can impede the complete interpretation and full utilization of the data so that clients do not achieve maximum benefit from their investments in the research undertakings. The second aspect of data utilization to be covered is the presentation of the results so clients can adequately understand the implications of the project and use the results accordingly in their decision-making processes.

Flexibility in Interpretation

Flexibility in the interpretation and analysis of the data does not mean data manipulation to achieve predetermined results. This defeats the purpose of a project undertaken to resolve a problem. Flexibility refers to researcher capability to really comprehend what the data and the analyses mean and how the relationships which they express can be used to advantage by the client in making a decision.

In close conformity to the needs and desires of the client, applied researchers should utilize all their training and experience as well as the knowledge gained from the current project in order to provide useful information. This includes a complete examination of the results to determine their meaning as well as their reliability. Over-emphasis on measures of reliability and insistence on rigid standards frequently set for more optimum conditions or different problems, reduce the ability of the researcher to explore the data in more detail and to fully understand the meaning of the results.

Meaning of the Results

A common fault in the research process is to accept results of the analyses as something sacred, even if they do not appear logical. An important aspect in the research process is the selection of appropriate theoretical concepts and choice criteria, the measures of performance, efficiency or success, which serve as guides in the construction of the problem statement. The performance criterion which should have been of concern to the researcher in our opening dialogue of this book was the increase in the total production of certain crops in his country for which trade agreements had just been made. An agronomist working on the development of a new variety may have as a performance criterion the resistance of the crop to a certain disease. The same agronomist also, of course, has a secondary criterion, that of increasing production per unit of land area in which the crop is grown. A farm economist usually considers the maximization of profit to some resource base as the most relevant performance criterion to use. Farmers may well have other criteria.

> Flexibility in interpretation refers to researcher capability to comprehend what the data and analyses mean and how the relationships expressed can help the client in decision making.

Blind adherence to a predetermined set of performance criteria and the lack of flexibility in considering alternatives can frequently obscure the real nature of the problem being treated. Three examples of errors in interpretation owing to the misuse or the misunderstanding of performance criteria will be discussed. These examples should provide the researcher some ideas of the kind of flexibility needed in the interpretation of research results.

Example 1.[4] The performance of potato producers in Colombia, as in most places, has been measured by research and extension specialists in tons per hectare simply because of professional tradition. Experimental yields per hectare improved annually from the early 1950s to the 1970s to the point where they at least tripled and often quadrupled average producer yields. Yet yields on both commercial and subsistence farms rose only slightly over the same period. The research and extension programs were often under criticism for not stimulating at least part of the yield increases known to be possible. At the same time, farm research revealed that both large and small producers were applying fertilizer and pesticides at levels near those recommended, and many were also using improved seed. Total potato production increased to keep pace with population growth mainly through the dedication of more land to the crop, but yields per hectare remained "embarrassingly" low. Should producers, agronomists, and extension specialists have been embarrassed? Maybe not.

Except for the few potato farmers who rent land for potatoes, land represented a relatively low cost input to even the small producers. The small farmers in the potato regions usually produced only this one crop as a cash crop. Hence, they considered all good land as potential potato land, and if feasible, they usually had marginal land that could also be put into potatoes. The farmers planted as much of their best available land as necessary to use their seed supply. It is this seed supply which is scarce, or an expensive input in the eyes of the producers. This is because during the year they had the alternatives of selling or eating the potatoes held in storage, some of which they were saving for seed. As the time for seeding got nearer, the price normally rose and the potatoes tended to spoil, creating even more temptation to sell the small amount of seed stock remaining. The purchase of seed, if necessary, created a real hardship on them and their families. Hence, at planting time producers had only a limited supply of good seed available. As a result, even if they had a small farm, they had more land available than seed to plant on it.

Thus, it was not yield per hectare which most interested the potato producers but yield per amount of seed used. They even reported yields

---

[4]Taken from Andrew, 1970.

in a yield-seed ratio such as 20:1. This goal (the farmers' performance criterion) implied wider spacing between rows and between plants than was recommended by agronomists who considered yield per hectare as a performance criterion. The result was a higher product-seed ratio, but a relatively low product-land ratio. Nevertheless, potato producers were reacting rationally to their resource situation and neither they nor their scientific advisor should have been embarrassed by low per hectare yields. In fact, the product-seed ratio had improved substantially in large part because of the technical assistance the producers were receiving. But at the same time, many research and extension workers remained frustrated because they were relying on an inappropriate, though traditional, performance criterion.

Example 2.[5] A similar situation occurred in West Pakistan where irrigation is necessary for crop production. Historically, water, whether provided by canals or by Persian wheel wells, was a much more limiting resource than land even though average farm size in many areas was only about seven acres. Contrary to recommendations, farmers were not using sufficient quantities of irrigation water to obtain highest yields per acre. Rather they spread the available water over more land so as to obtain maximum production per unit of water (they still were not able to use all their land at any one time). Before more water was made available, an increase in water application per acre would have reduced the amount of land they could irrigate and would have reduced the crop-water ratio, resulting in less total crop for the very limited quantity of water. The gain in yield per acre clearly was not an appropriate goal. Of course with more water and the use of new technical inputs, increased production per acre could become a relevant goal.

Example 3.[6] Returning to Colombia, some questions arose concerning productivity measures for dairy farms. Performance of a dairy herd is traditionally measured by the average pounds of milk produced annually per cow for the entire herd, including dry cows. The goal of a dairy improvement program is usually to maximize this

---

[5]Experience by Peter Hildebrand on an assignment with Tipton and Kalmbach, Inc., Engineers in West Pakistan, 1964-1966, and reported in Tipton and Kalmbach.

[6]Unpublished research by Rex Rehnberg and L. C. Garrison, Universidad Nacional - Instituto Colombian Agropecuario, Medellin, Colombia, 1970.

average. This performance criterion requires that breeding programs develop large cows with well developed udders and other characteristics conducive to high yields per animal. It was assumed in Colombia that protein was lacking in the rations for dairy cattle.

Preliminary research suggested, however, that the very hilly, mountainous and rainy pasture conditions found in Colombia may have required both different nutrition and breeding research programs. The large dairy cow could not move about in the mud and on the hills nearly as well as a small one. Maybe three small cows could better adapt to these conditions than two large cows while better utilizing the same amount of pasture even though yield per cow would be sacrificed. Of course milking costs, veterinary costs and so on for alternative herd sizes should be considered. As to nutrition, the energy requirements for the cows appeared to exceed most expectations making energy rather than protein the most important deficiency in the usual ration. These preliminary conclusions resulted because of a flexible attitude on the part of the researchers in interpreting data which initially appeared not to be very logical.

Reliability of Results

The degree of reliability of empirically based research results is nearly always assessed by statistical tests of significance which utilize confidence limits or levels to judge how much reliability one may put in the results. A common research practice is to set confidence limits before data collection is undertaken and then to reject results which cannot be measured to that fixed degree of precision. For example, a null hypothesis is accepted if the tests of significance that are used reach a certain level of confidence and rejected if that level is not reached. Terms in an equation fit by regression analysis are accepted at certain levels of significance and rejected below these fixed levels. In analysis of variance, differences in responses to various treatments must meet certain levels of significance before it is said that the differences are real and significant.

For some research projects, however, significance tests may be impossible and even unnecessary depending upon the nature of the problem and resources available to resolve it. McPherson notes, for example, that economists, when adjusting experimental data for predicting a production function, can rely upon the experience of specialists in several basic science disciplines for knowledge about unmeasured environmental and agronomic conditions which influence crop production. "The fact that 'expert' estimates would not be subject to statistical tests of significance does not mean necessarily that such estimates for particular purposes would be less accurate than estimates derived from sources that lend themselves to statistical tests." [McPherson, p. 804].

In other research projects, strict adherence to predetermined levels of significance is necessary in order to preserve objectivity and protect the client from unnecessary risk. Experimentation with medicines is a common example where very high significance levels must be maintained if the medicines are for human use.

But there is also a great deal of research, particularly applied research, where adherence to the strict observation of high significance levels can unduly restrict the amount of information made available to the client, and can, in fact, be detrimental to the client's best interests. Consider, for example, the case of a low cost practice which appears to increase the yield of a particular crop, but the increase in yield is not significant at the 90 percent or 95 percent levels of confidence. If we tell our clients that there was no significant increase in yield, they will probably decide against its use. But if the cost of the practice is low, then the financial risk to the producer is also low (assuming that there is no reduction in yield due to the use of the practice). It might well be that producers are willing to accept this risk if they have even a 50/50 chance of getting an increase in yield that would be worth 10 times their investment if, in fact, there really was an increase. In this case, would it not be better for the client to receive information saying that there was an increase in yield in most cases with the use of the practice, but it requires a reduction in confidence limit to, say, 70 percent to be a statistically significant increase?

A 70 percent confidence level, in essence, means that if this same experiment were repeated 10 times under the same conditions chances are that at least 7 of the 10 experiments would indicate an increase in yield. Note that neither a 70 percent nor a 95 percent confidence level tell us that there will or will not be an increase in yield on any particular farm. That effect is only implied by conducting the experiment under conditions similar to those of the client or clients. But if there is a valid similarity between the experiment and the conditions under which the client will use the practice, then the costs and the probable returns of the practice to the producer (the economics of the practice) will ultimately determine if the clients will utilize that particular practice on their farms.

In applied research, it is important to remember that the client is waiting for conclusions and recommendations. If the research process has been thorough, then no additional information will be available to help the researcher in making a conclusion unless additional experimentation is conducted. If the researcher insists on additional research before he or she is willing to make any conclusions and recommendations then it will mean a delay before the farmer or other client can put the practice into effect. This represents potential profit foregone by the client (if in fact a change would have been profitable) and should definitely be considered by the researcher and by the client before making a decision not to draw any conclusions from the results of the completed experiment. When necessary and appropriate, one can utilize sensitivity analysis to determine the economic costs of making the wrong decision [Ackroff, pp. 39-45; Havlicek and Seagraves; McPherson and Farris].

> The degree of reliability measured by statistical levels of significance applied to the results must be specified relative to the client's need and the nature of the problem.

Although the decision to repeat the experiment may mean a delay of only one year to the client, the economic consequences as well as the decision not to draw a conclusion are related to the so called Type I and Type II errors which can be made in accepting and rejecting hypotheses. A hypothesis is stated in such a form that it is either true or false. The research process generates the evidence upon which the researcher will decide whether or not to accept the hypothesis as stated. But any acceptance or rejection of a meaningful hypothesis is always associated with a probability of making an error (it is virtually impossible to be 100 percent sure that a hypothesis is true or that it is false). A Type I error is committed when a hypothesis is rejected (thought by the researcher to be false) when, in fact, it really is true. If there truly is a response to a practice and the research hypothesis states that there is, then the researcher is committing a Type I error if he or she advised the client that there is no response because the research indicated the hypothesis was false. Note that the same consequences follow if the researcher, because of adherence to a too high confidence level is unwilling to decide and hence advises the client that it is not possible to determine if there is a response.

In this same example, the researcher is committing a Type II error if he or she accepts the hypothesis (thinking it to be true) when, in fact, it is false. In this case they advise the client that the practice results in an increased yield when, in fact, it does not. The consequences to the client of the two types of error are obviously different.

If the client follows the conclusions (and presumably the recommendations) of the researcher then in the first case (Type I error), the client does not adopt the practice which would have increased yield and forgoes potential income which would have been earned if an investment had been made in the practice. In the case of a Type II error, the client invests cash in the practice without results and loses the money invested. The Type II error can be even more serious if the practice actually reduces yields (the true effect is negative and not zero).

There is an important relationship between the two types of errors. A change in the confidence level in an attempt to reduce one type of error increases the probability of making the other type of error. Too much concern with the Type II error (the one most commonly considered) can raise confidence levels so high that a Type I error becomes highly probable. When researchers want to be very sure that there is a positive response before they recommend a practice, they may, in effect, be advising their client that there is no response even though one does exist. This is usually the case when the only recommendation made from an experiment is that it should be repeated with better control, or that a survey must be repeated with a larger sample or better questionnaire.

The need for flexibility in applied research should be obvious from the above examples. A researcher may not need to involve the client in decisions about specific tests once the general confidence guidelines have been specified. But a good dialogue throughout the data interpretation process may be important if new questions arise concerning the overall confidence desired. For some problems which may be extremely difficult to research or for specific hypotheses, the client may be willing to accept a recommendation which should hold true only a majority of the time. Other issues may be more critical and require very high levels of confidence. But neither the client nor the researcher should blindly insist on high confidence levels because to do so may result in the loss of a great deal of useful information.

Finally, the end result of an applied research program may not with any degree of confidence resolve the problem to which it was directed. Lest this statement seem completely out of context after all that has been said in the book, we should hasten to add that few problems are totally resolved. In many cases, the end result of research is the identification of other related problems and possibly hypotheses concerning their resolution. Often, when reporting applied research results, definitive recommendations which may help resolve the problem are accompanied by a new question or hypothesis. This is true because applied research is a continuous quest for knowledge which appears to have few bounds beyond time and resource restrictions.

Presentation of the Results

By now the need for drawing conclusions from applied research and making recommendations to the client should be obvious to the reader. The researcher is the expert to whom the client is looking for help. If researchers are unwilling to fulfill their role, there is little justification for their having undertaken a project.

> The final activity in applied research is the communication of the results to the client.

We have encouraged flexibility and the use of the researcher's experience and his or her opinions in the interpretation of the data and in the analyses. This is important to the client. But the client is also entitled to a clear exposition as to what part of the conclusions result from the conjecture of the researcher and what part is attributable to the pure analysis of the data. As a scientist, the researcher is obligated to report the results of the research project honestly and objectively. As a professional hired to provide information for resolving problems, the researcher is also obligated to interpret these results in accordance with the needs of the client.

The final stage in the applied research process is the communication of the results to the client. It is as essential that the client comprehend what the researcher has to say as it is for the researcher to understand the nature of the client's problem and the meaning of the research data. The researcher is a scientist whose job it is to understand complicated analyses and confusing data. The client has other interests and ordinarily does not have the same training, so the researcher must report findings in a language that the client will understand.

In many cases it is useful to clients to have the results presented in more than one form. They are entitled to, and should receive, a complete report on the project including a review of the problem as finally defined, the hypotheses and the objectives. These all appear in the final report much as they were written in the project proposal. In discussing the conclusions, it is necessary to refer to the hypotheses and show which were accepted and which rejected and how each affected the final conclusions. A clear exposition of this part of the research process provides the client with the basis for understanding the recommendations made by the researcher.

> The client expects to have a report on the project including a review of the problem, hypotheses, objectives and results.

The extent to which the researcher should detail the analyses and data collection procedures, the more technical aspects of the research, depends on the familiarity of the client with these techniques and the client's ability to understand them. Frequently, the researcher can make one technical report and another briefer report can be written in non-scientific language, excluding the technical details. The detailed report can be used by the client or another professional to verify the validity of the research procedures should the need arise or to serve as a guide to further research, while the shorter report will usually be more functional for the client to use in the decision-making process.

The majority of applied research in agriculture serves farmers as the principal clients through the activities of an extension service.[7] Many farmers are not interested in how the recommendations came to be made but they need to know what the recommendations and alternatives entail and what they may mean if adopted. Many farmers depend on the researcher and the extension specialist or agent to interpret the results of the research process. In this case it is most useful to prepare a research report which presents only the specific recommendations and the nature of the effect or response which can be expected if the recommendations are followed. An excellent procedure is for the researcher to present results in a form directed toward extension specialists and agents and then work with the extension personnel to prepare a non-technical publication for use in the extension programs.

---

[7]Research, extension and education can and must blend together in addressing current and future problems. The need and approach are documented in many sources including Boehlje, Eidman and Walker; Caunder; Farrell; Guither; Holt, Pugh and Brant; Shaffer.

> Tell the client what the alternatives and recommendations entail and what they may mean if adopted.

A thesis or dissertation is a research report whose language and content are directed necessarily toward other researchers -- primarily the student's supervisory committee. Few theses and dissertations are usable in their entirety as a research report for more popular distribution, but they can serve as a basis for the generation of a series of documents, each directed toward a different client or audience. Articles for technical journals can report the factors of scientific interest which occurred in the research process. One or more popular reports or bulletins can be directed toward the ultimate clients or intermediate groups such as the extension service. Finally, leaflet type publications which present the recommendations in simplified form can also be prepared if the research is applicable to a wider audience.

Regardless of who the clients are, the need to present them with the results of applied research cannot be overemphasized. Too often, especially in institutional research where the client tends to be an impersonal group not known directly by the researcher, the results of completed projects are not prepared in a form which is useful to the client. When this occurs, the client can become disenchanted with the research institute and financial difficulties often result. Hence, one of the best ways to assure a continuing source of research funds and a demand for research services is to present the client with a useful research product.

Summary

The culmination of the applied research project is the presentation of the results to the client. The form in which they are presented will depend on the ability of the client to understand the nature of the research process. Many clients will be concerned only with the recommendations of the researcher and what they can mean to them if they were to follow what is recommended. Other clients will desire a complete research report including a review of the problem being studied, the hypotheses and the objectives of the project. The sophisticated client will also want information on the data collecting procedure and the analyses which were used.

Even the sophisticated client depends on the researcher for an interpretation of the results and will want to consider the researcher's recommendations in the decision-making process. At the termination of the project, the researchers should be considered as experts in the subject and they should use their knowledge and experience accordingly in interpreting the results for their clients, who expect the full utilization of the resources invested in the project or program.

Perhaps the most important conditions for your success as an applied researcher is implementation of a methodology that provides for careful *identification of the client's problem.*

In this book, we have described the process of planning and executing applied research undertaken for the resolution of a specific problem. Planning includes the specification of the problem in such a form that it is researchable within the resource limitations facing the researcher, the formulation of testable hypotheses which suggest meaningful solutions to the problem, and the delineation of objectives which describe what is expected to be achieved by the project. The execution of the project includes the collection and analysis of data, the drawing of conclusions, and the making of recommendations.

Throughout the book, the justification for applied research is considered from the point of view of service to a client. The client's need to make a decision imposes a deadline on such research and the magnitude and importance of the problem, along with the nature of the research to be conducted, determine the quantity of resources required for the project and their availablility to the researcher. Within these time and resource limitations, the researcher must strive to create a product useful to the client. With this book, we present researchers with an approach to research under these limiting conditions which will help them to become efficient, effective and of greater service to their clients.

A Recap

- Research begins and ends with management of one's creative self.
    - To that end a prepared mind for research draws on conscious and subconscious potentials.

- Managing creative research involves an ability to work with diversity, complementarity, synthesis and consensus.
- You have an obligation to understand the moral and ethical dimensions of your research.

- Problem oriented research is a process of deduction based in theory.
  - Humans naturally develop and test theory from observed or experienced relationships.
  - Theory provides structure and form to "known" relationships while extending the learning process and the knowledge base.
  - Theory formulation and application is initiated with the hypothesis.

- Perhaps the most critical deficiency in applied research methodology is the failure to identify the specific problem.
  - The most serious consequence of weak applied research is the cost associated with not having relevant information for making decisions.
  - An effective applied research methodology is directed toward the efficient use of available research resources to maximize the probability of achieving meaningful results to help resolve problems.

- Decisions are necessary and applied researchers are expected to provide useful information even within limited research resource situations.
  - Applied research is accomplished under time, financial, human; information and physical resource constraints.
  - A plan should give a high probability of accomplishing research to help resolve the problem within the research resource constaints.
  - The urgency of problem resolution in applied research makes time an important resource.
  - Data and information requirements are narrowed and brought into focus by careful research planning.
  - The methodology, budget and calendar are interdependent parts of the whole research proposal and resource plan.

- Open-ended approaches can assist with problem identification for research planning or provide early information when the time constraint is more immediate.
  - Open-ended approaches to information gathering capture qualitative responses through participation and dialogue with informants and respondents in case studies, focus groups, multi-disciplinary teams and cooperative inquiry teams.
  - Focus groups provide qualitative information through focused discussion by a relatively homogeneous group of people.

- The multidisciplinary research team obtains information from informants as a basis for holistic responses to client needs.
- Cooperative inquiry is collaborative, creative and directed learning (inquiry) by researchers and people involved in the situation under study. It expands the range of participatory research.
- As a research tool, the case study provides a detailed and holistic process for observing the structure and flow of real-life events.
- Use of non-experimental secondary data requires researcher responsibility in verifying the quality and applicability of the data to the problem under study.

- Careful development of the problem statement, the hypotheses and the objectives is the key to efficient research and the most productive use of time by the researcher.
  - A well conceived plan will increase the probability of accomplishing research to help resolve the problem.
  - An unmanageable research project yields few benefits to anyone.
  - Few people write an acceptable research proposal on the first attempt.
  - There is no single right, or correct researchable problem -- there is a correct framework for expressing the problem for useful research.
- A problem exists when there is a need felt by a client.
  - Hypotheses are tentative propositions that must relate to and help resolve the problem.
  - Objectives specify what the researcher intends to do and what information will be obtained for the client.
- The advantage of experimental data over non-experimental data is the degree of control held by the researcher over the variables under study.
  - Human subject interaction in research varies depending on the degrees to which bio-physical, experimental and behavioral approaches are combined and utilized.
  - No easy lines are drawn between client needs for descriptive, predictive and prescriptive research.
- Research planning is the basis for choosing an experimental design.
  - Precision, accuracy, relationship to the problem and efficiency must be considered in the choice of experimental design.
  - In determining relationships to assist with problem resolution, the design of a useful experiment must be related to the problem.

- Size and complexity of an experiment depend on many factors; ultimately time and resource constraints encourage the use of an effective research plan and experimental design.
- The nature of the problem under study suggests the statistical precision required in the experimental results.
- Experimental data generated by another person from other experiments must be used with caution to assure they are relevant and comparable.
- Some research problems, settings and resource situations require multi-disciplinary, multi-purpose and multi-locational experimentation where farms serve as replications and/or environments within an on-farm research orientation.
- Multidisciplinary experimentation is a means of efficiently using scarce research resources.
- In non-experimental research, the researcher usually cannot determine design and level of the variables nor directly measure the phenomena, but controls only the measurement technique.
- Flexible, applied, non-experimental researchers place emphasis on the research problem and view their sampling techniques as tools and not ends.
- For effective research resource use, questionnaire design and implementation must deal with relevancy, language, cross-cultural credibility, measurement and other challenges.
- No amount of intellectual exercise in the office can substitute for properly testing a questionnaire and field training of interviewers.
- Verification is necessary so that the researcher can be confident that the data best measure the phenomena needed to fulfill the research objectives.
- Finally when the moment of truth arrives, the client is expecting a useful product from your information gathering, analysis and interpretation. Now YOU are the most knowledgeable person available to the client and because the client must make a decision your product must be usable and understandable.
- Flexibility in interpretation refers to researcher capability to comprehend what the data and analyses mean and how the relationships expressed can help the client in decision-making.
- The degree of reliability measured by statistical levels of significance applied to the results must be specified relative to the client's need and the nature of the problem.

- The final activity in applied research is communication of the results to the client.
- The client expects to have a report on the project including a review of the problem, hypotheses, objectives and results.
- Tell the client what the alternatives and recommendations entail and what they may mean if adopted.

- Perhaps the most important condition for your success as an applied researcher is implementation of a methodology that provides for careful *identification of the client's problem.*

References

Ackroff, R.L. 1953. The Design of Social Research, The University of Chicago Press.

Allin, B.W. 1949. Theory: Definition and Purpose, Journal of Farm Economics. XXXI: 3. pp. 409-417.

Andrew, C.O. 1969. Improving Performance of the Production-Distribution System for Potatoes in Colombia. Ph.D. dissertation, Michigan State University.

Andrew, C.O. 1970. Problemas en la Modernizacion del Proceso de Produccion de Papa en Colombia. Departamento de Economia Agrícola, Instituto Colombiano Agropecuario, Boletín Departamental No. 6.

Arnon, I. 1968. Organization and Administration of Agricultural Research. Elsevier Publishing Co., Ltd., Amsterdam, London, New York.

Ayer, H.W. and E.E. Schuh. 1972. Social Rates of Return and Other Aspects of Agricultural Research: The Case of Cotton Research in Sao Paulo, Brazil. American Journal of Agricultural Economics. 54:4. Part I. pp. 557-559.

Back, W.B. 1961. Philosophy, Methods and Status of Agricultural Economics. Journal of Farm Economics. XLIII:4. Part I. pp. 898-909.

Baum, E.L., et. al. (eds.). 1957. Economic and Technical Analysis of Fertilizer Innovations and Resource Use. Iowa State University Press, Ames.

Baum, E.L., E.O. Heady and J. Blackmore (eds.). 1956. Methodological Procedures in the Economic Analysis of Fertilizer Use Data. Iowa State University Press, Ames.

Bear, D.V.T. and D. Orr. 1967. Logic and Expediency in Economic Theorizing. The Journal of Political Economy. 75:2. pp. 188-196.

Boehlje, M., V. Eidman and O. Walker. 1973. An Approach to Farm Management Education. American Journal of Agricultural Economics. 55:2. pp. 192-197.

Boulding, K. 1956. The Image. University of Michigan Press, Ann Arbor. 1956.

Brandow, G.E. 1955. Methodological Problems in Agricultural Policy Research. Journal of Farm Economics. XXXVII:5. pp. 1316-1324.

Brauer, J. and A. Hunter. 1989. Multimethod Research, A Synthesis of Styles. Sage Library of Social Research 175. Sage Publication, Newbury Park, California.

Brownlee, K.A. 1965. Statistical Theory and Methodology. John Wiley & Sons, Inc., New York, London, Sydney.

Buse, R.C. 1973. Increasing Response Rates in Mailed Questionnaires, American Journal of Economics. 55:3. pp. 503-508.

Caunder, R. 1969. Identifying Extensions Community Resource Development Clientel and Adapting Economic Information to the Needs of Each. Southern Journal of Agricultural Economics. 1:1. pp. 109-112.

Christensen, R.L. 1966. Logic, Science, and Economics. Journal of Farm Economics. 48:1. pp 137-139.

Cochran, W.G. 1963a. Experimental Design. John Wiley & Sons, Inc., New York, London, Sydney.

Cochran, W.G. 1963b. Sampling Techniques. John Wiley & Sons, Inc., New York, London, Sydney.

Cohen, M.R. and E. Nagel. 1934. An Introduction to Logic and Scientific Method. Harcourt Brace & Co.

Conant, J.B. 1951. On Understanding Science: An Historical Approach. New York-New American Library, New York.

Conant, J.B. 1951. Science and Common Sense. New Haven, Yale University Press.

de Bono, E. 1970. Lateral Thinking, Creativity Step by Step. Harper and Row.

Dewey, J. 1938. Logic, The Theory of Inquiry. New York, H. Holt & Company.

Dillman, D.A. 1989. Mail and Telephone Surveys. John Wiley & Sons, New York.

Escobar, G., et. al. 1971. La Ceba de Novillos en Confinamiento. Departamento de Economía Agrícola, Instituto Colombiano Agropecuario. Informe No. 6.

Ezekiel, M. and K.A. Fox. 1963. Methods of Correlation and Regression Analysis. John Wiley & Sons, Inc., New York, London.

Farrell, K.R. 1971. New Priorities in Agricultural Research: Implications for Economists in the West-An Extension View. Western Agricultural Economics Association. Proceedings, 44th Annual Meeting. pp. 10-12.

Feaster, G.W. 1970. An Analysis of the Relationship Between Infrastructure and Agricultural Development in Caqueta, Colombia. Departamento de Economía Agrícola, Instituto Colombiano Agropecuario. Boletín Departamental No. 14.

Feldstein, H.S. and S.V. Poats, Eds. 1989. Working Together: Gender Analysis in Agriculture. Vol. 1: Case Studies. Vol. 2: Teaching Notes. Kumarian Press, Inc.

Friedman, M. 1953. Essays in Positive Economics, Part I. University of Chicago Press.

Gallo, A. 1971. Analisis Economico de los Problemas en la Produccion de Cerdos en Anitoquia, Antiguo Caldas y Valle. Unpublished M.S. Thesis, Universidad Nacional -- Instituto Colombiano Agropecuario, Tibaitata, Colombia.

Gallo, A. and Hildebrand, P.E. 1971. Analisis Económico de la Ceba de Novillos en Pastoreo Rotacional. Departamento de Economía Agrícola, Instituto Colombiano Agropecuario. Informe No. 5. Regional No. 5.

Gibson, W.L. Jr., R.J. Hildreth and G. Wunderlich, (eds.). 1966. Methods for Land Economics Research. University of Nebraska Press, Lincoln.

Good, W.J. and P.K. Hatt. 1952. Methods in Social Research. McGraw Hill Book Company, Inc.

Guither, H.D. 1973. Extension Education in Bargaining Among Midwest Livestock and Grain Producers. American Journal of Agricultural Economics. 55:1. pp. 58-60.

Hall, E.T. 1959. The Silent Language. A Premier, Doubleday & Company.

Halter, A.N. and H.H. Jack. 1961. Toward a Philosophy of Science for Agricultural Economic Research. Journal of Farm Economics. XLIII:1. pp. 83-95.

Halter, A.N. 1960. The Identification of Problems in Agricultural Economics. Journal of Farm Economics. XLIII:5. pp. 1459-1471.

Hansen, M.H., W.N. Hurwitz and W.G. Madow. 1953. Sample Survey Methods and Theory. Volumes I & II. John Wiley & Sons, Inc., New York, London.

Hausman, D.M. 1984. The Philosophy of Economics -- An Anthology. Cambridge University Press.

Havlicek, J. and J.A. Seagraves. 1962. The Cost of the Wrong Decision as a Guide in Production Research. Journal of Farm Economics. XLIV:1. pp. 157-168.

Hildebrand, P.E. 1972. Análisis Agroeconómicos Mediante Superficies de Respuesta. Ministerio de Agricultura y Ganadería, Dirección General de Economía Agrícola y Planificación, Departamento de Administración Agricóla, San Salvador, El Salvador.

Hildebrand, P.E. 1981a. Combining Disciplines in Rapid Appraisal: The Sondeo Approach. Agricultural/Administration. 8: pp. 423-432.

Hildebrand, P.E. 1981b. Motivating Small Farmers, Scientists and Technicians to Accept Change. Agricultural Administration. 8: pp. 375-383.

Hildebrand, P.E. 1984. Modified Stability Analysis of Farmer Managed, On-Farm Trials. Agronomy Journal 76: pp 271-274.

Hildebrand, P.E. and F. Poey. 1985. On-Farm Agronomic Trials in Farming Systems Research and Extension. Lynne Rienner Publishers, Inc. Boulder, Colorado.

Hildreth, R.J. and Castle, E.N. 1966. Identification of Problems, in Methods for Land Economics Research, edited by: W.L. Gibson, Jr., R.J. Hildreth and Gene Wunderlich. University of Nebraska Press, Lincoln.

Holt, J., C.R. Pugh and W.L. Brant. 1973. Educational Programs for Commercial Agriculture and Agribusiness, Southern Journal of Agricultural Economics. 5:1. pp. 47-54; and discussion by J.M. Sprott, pp. 55-57.

Jhunjhunwala, Bharat. 1973. Agricultural Mechanization, Rural Income Distribution and Unemployment in Faizabad District, East Uttar Pradesh. Unpublished Ph.D. dissertation, University of Florida.

Johnson, G.L. 1986. Research Methodology for Economists, MacMillan, New York.

Kant, I. 1783. Prolegomena to Any Future Metaphysics. Recent publication in 1950 by the Liberal Arts Press, Inc. New York.

Kaplan, A. 1964. The conduct of Inquiry. Chandler Publishing Co.

Kelso, M.M. 1965. A Critical Appraisal of Agricultural Economics in the Mid-Sixties. Journal of Farm Economics. 47:1. pp. 1-16; and comments including the following from the Journal of Farm Economics. 47:3. 1965. pp.843-857: W.G. Brown. An appraisal of a Critical Appraisal; W.J. Donova. Current Status of Agricultural Economics as a Science; E.W. Grove. Basic Research in a Quasi Science; Ronald L. Mighell. Is Agricultural Economics; R.G.F. Sptize. Agricultural Economics: Predictive or Prescriptive. M.M. Kelso. The Authors Last Word -- For Now.

Kiehl, E.R. 1957. Integration of the Sciences for Effective Research. Journal of Farm Economics, XXXIX:5. pp. 1230-1231.

Kmenta, J. 1971. Elements of Econometrics. The MacMillan Company, New York.

Knoblauch, H.C., E.M. Low and W.P. Myer. 1962. State Agricultural Experiment Stations: A History of Research Policy and Procedure. USDA Misc. Pub. 904.

Krueger, R.A. 1988. Focus Groups, A Practical Guide for Applied Research. Sage Publication, Newbury Park, California.

Krupp, S. 1963. Analytic Economics and the Logic of External Effects. American Economic Review. LIII:2. pp. 220-226.

Kuznets, G.M. 1963. Theory and Quantitative Research. Journal of Farm Economics. 45:5. pp. 1393-1403.

Ladd, G.W. 1987. Imagination in Research, An Economists View. Iowa State University Press.

Leontief, W. 1966. Essays in Economics: Theories and Theorizing. Oxford University Press.

Manderschied, L.V. 1964. Some Observations in Interpreting Measured Elasticities. Journal of Farm Economics. ILVI:1. pp. 128-136.

Manley, W.T. 1969. An Essay on Federal-State Research Programs in Agricultural Economics: Need and Prospects for the Future in Agricultural Marketing, Southern Journal of Agricultural Economics. I:1. pp. 119-122.

Mann, P.H. 1968. Methods of Sociological Enquiry. Schocken Books, New York.

Marshall, C. and G.B. Rossman. 1989. Designing Qualitative Research. Sage Publications, Newbury Park, California.

McDermott, J.K. and C.O. Andrew. Management of Agricultural Research and Extension (course notes to be published).

McKee, D.E. 1972. Our Obsolete Data Systems. American Journal of Agricultural Economics. 54:5. pp. 867-880.

McPherson, W.W. 1952. Sources and Use of Technical Data in Farm Management Analysis. Journal of Farm Economics. XI:4. pp. 803 and 804.

McPherson, W.W. and J.E. Farris. 1958. 'Price Mapping' of Optimum Changes in Enterprises. Journal of Farm Economics. XL:4. pp. 821-834.

Miller, D.C. 1964. Handbook of Research Design and Social Measurement. David McKay Company, Inc.

Morgon, G. 1983. Beyond Method, Strategies for Social Science Research. Sage Publications, Newbury Park, California.

Motes, W.W. 1972. Economic Research Programs for Community Development-Why Do It?. Southern Journal of Agricultural Economics. 4:1. pp. 9-13.

Murphy, J. 1985. The Power of Your Subconscious Mind. Bantam Books.

Nagel, E. 1961. The Structure of Science. Harcourt, Brace and World.

Nagel, E. 1963. Assumptions in Economic Theory. American Economic Review. LIII:2. pp. 211-219.

National Research Council. 1989. Alternative Agriculture. Committee on the Role of Alternative Farming Methods in Modern Production Agriculture, Board on Agriculture, NRC, National Academy Press. Washington, D.C.

Noltingk, B.E. 1965. The Art of Research, A Guide for the Graduate. Elsevier Publishing Company, Ltd., Amsterdam, London, New York.

Paarlberg, D. 1963. Methodology for What?. Journal of Farm Economics. 45:5. pp. 1386-1392.

Paarlberg, D. 1969. Norman Borlaugh-Hunger Fighter. Foreign Economic Development Service, U.S. Department of Agriculture and U.S. Agency for International Development. PA 969.

Parsons, K.H. 1949. The Logical Foundations of Economic Research. Journal of Farm Economics. XXXI:4. pp. 656-686.

Pearson, K. 1937. The Grammar of Science. London, J.M. Dent & Son, Ltd.

Peirce, J.R. 1954. The Social Uses of Science. American Scientist. October. pp. 495-497.

Peng, K.C. 1967. The Design and Analysis of Scientific Experiments. Addison-Wesley Publishing Co., Reading, Massachusetts.

Plaxico, J.S. 1970. Agricultural Development Effort Assessment: The Colombian Case. Southern Journal of Agricultural Economics. 2:1. pp. 69-76.

Reason, P. 1988. Human Inquiry on Action, Developments in New Paradigm Research. Sage Publications, Newbury Park, California.

Rogers, E.M. 1964. Diffusion of Innovation. The Free Press of Glencoe and MacMillan, New York and London. pp. 7-12.

Rose, E.M. 1948. Selection of Problems for Research. American Journal of Sociology. 54. pp. 219-229.

Runes, D. 1960.. Dictionary of Philosophy. Philosophical Library, Inc.

Ruttan, V.W. 1982. Agricultural Research Policy. University of Minnesota Press.

Salter, L.A. 1967. A Critical Review of Research in Land Economics. The University of Wisconsin Press, Madison.

Schermerhorn, R.W. 1969. Identifying Extension's Marketing Clientele and Adapting Economic Information to their Needs. Southern Journal of Agricultural Economics. 51:1. pp. 113-118.

Schuh, G.E. 1970. Research on Agricultural Development in Brazil. The Agricultural Development Council, Inc., New York, N.Y.

Shaffer, J.D. 1965. Some Conceptual Problems in Research on Market Regulations, in Federal, State and Local Laws and Regulations Affecting Marketing. Regional Res. Bul. No. 168. North Dakota State University.

Singh, B.K. 1990. Sustaining Crop Phosphorous Nutrition of Highly Leached Oxisols of the Amazon Basin in Brazil through Use of Organic Amendments. Ph.D. Dissertation, University of Florida.

Skjonsberg, Else. 1989. Change in an African Village: Kefa Speaks. Kumarian Press, Inc.

Smith, E.D. 1970. University Assistance in Building Agricultural Research Institutions in Southeast Asia - A Case Analysis. Southern Journal of Agricultural Economics. 2:1. pp. 61-67.

Steel, R.G.O. and James H. Torrie. 1980. Principles and Procedures of Statistics: A Biometrical Approach. Second Ed. McGraw-Hill Book Company.

Stern, A. 1956. Science and the Philosopher. American Scientist. July.

Sudman, S. 1967. Reducing the Cost of Surveys. Chicago Aldine Publishing Co.

Sundquist, W.B. 1969. Federal -- State Research Programs in the Economics of Agricultural Production -- Needs and Prospects for the Future. Southern Journal of Agricultural Economics. 1:1. pp. 123-127.

Taylor, J. 1983. Dream Work, Techniques for Discovering the Creative Power in Dreams. Paulist Press, New York.

Theil, H. 1971. Principles of Econometrics. John Wiley & Sons, Inc., New York, London, Sydney, Toronto.

Thirring, H. 1956. The Steps From Knowledge to Wisdom. American Scientist. October. pp. 445-446.

Tipton and Kalmbach. 1967. Regional Plan, Northern Indus Basin-Development and Use of the Water Resources of the Indus Basin. Volume II. Economics, West Pakistan Water and Power Development Authority and Tipton and Kalmbach, Inc., Engineers. 1967.

Tripp, Robert, Ed. 1991. Planned Change in Farming Systems: Progress in On-Farm Research. John Wiley and Sons.

Vining, R. 1960. Methodological Issues in Quantitative Economics: Variations Upon a Theme by F.H. Knight. American Economics Review. XL:3. pp. 267-284.

Walsh, W.H. 1967. Kant, Immanuel. The Encyclopedia of Philosophy. Vols. 3 & 4. Paul Edwards, Editor in Chief. Macmillan Publishing Co., Inc. and the Free Press, New York. pp. 305-325.

Weaver, W. 1961. The Imperfections of Science. American Scientist. 49:1. pp. 99-113.

West, Q.M. 1973. The Changing Mission of Agricultural Economics Research. Division of Information, Economic Research Service, USDA. June.

White, J.H. 1969. Relevant Research Areas and Organizational Questions Relative to Federal-State Research Programs in the Economics of Agricultural Production. Southern Journal of Agricultural Economics. 1:1. pp. 129-132.

Whyte, W.F. ed. 1991. Participatory Action Research. Sage Publications. Newbury Park, California.

Wilson, C.P. 1971. New Priorities in Agricultural Research -- Implications for Economists in the West: An Administrative View. Western Agricultural Economists Association. Proceedings, 44th Annual Meeting. pp. 106.

Yee, L.K. 1991. A Process For Incorporating Social and Environmental Concerns in Food Safety Discussions. Speech in the November Conference on Agricultural Systems: Incorporating Environmental and Social Concerns. University of California Davis.

Yin, R.K. 1989. Case Study Research Design and Methods. Sage Publications. Newbury Park, California.

Zetterberg, H.L. 1965. On Theory and Verification in Sociology, The Bedminister Press.

Zimet, D.J. 1974. The Economic Potential for Increasing Vegetable Production in the Zapotitan District, El Salvador. M.S. thesis, University of Florida.

APPENDIX:

Applied Research Project Proposals

Each of the following five proposals provides a case for use by the student of applied research methodology. All of the proposals conform generally to the approach of the book; yet in reviewing the proposals one finds diversity both in the content and form of the problem statements, research hypotheses and research objectives. Our minor editing has purposefully not demanded detailed conformity by the authors to our dispositions. Instead, the proposals serve as study cases. All proposals stand to be improved and we challenge you to review and use these proposals with that thought in mind. We express gratitude to the student authors of the proposals and to the many students who might too have featured their proposals in the book. In no way do these selections pretend to represent the wide range of topics and problems encountered by applied agricultural researchers.

APPENDIX

Applied Research Proposals [illegible]

Case One
MARKET ORGANIZATION

I. Project Title: The Florida Shrimp Processing Industry Economic Structure and Market Channels

II. Project Leader: Jose Alvarez, Agricultural Economist

III. Anticipated Duration: January 1, 1973 to December 31, 1973.

IV. Problem Statement:

A. Problematic Situation:

In 1970, Florida produced over $112 million of processed seafood products and was the leading state in the Southeast Region. Shrimp products, valued at $262 million, accounted for over 63 percent of the value of all processed seafood products in the Southeast Region in 1970. That year, Florida accounted for over 30 percent of all shrimp products processed in the Region.

Shrimp is the most important seafood processed in Florida. Total value of the shrimp processed in the State in 1970 was slightly over $79 million or over 70 percent of the value of non-industrial seafood products.

Florida's shrimp processing industry has been growing in both absolute and relative magnitudes during the last decade. The State processed about 34 million pounds of shrimp in 1960 which were valued at more than $25 million. More than 68 million pounds were processed at a value of $79 million in 1970. While Florida accounted for 19 percent of the total volume of processed shrimp consumed in the U.S. in 1960 and for 21 percent of value, both percentages changed to 25 by 1970.

Even though growth of Florida's shrimp processing industry during the last decade was substantial, shrimp landings in the state show a sharp decline from over 51 million pounds in 1960 to 29 million pounds in 1973. Only in 1960 and 1961 were landings greater than quantity processed. By 1970 Florida processed almost 96 million pounds and landed slightly over 31 million, representing a deficit of over 64 million pounds to be filled primarily by imports. Competition for imports, particularly with Japan and Canada, is becoming increasingly intense.

The importance of Florida in the U.S. shrimp processing industry coupled with the growth of U.S. imports as a source of raw product for processors suggests the importance of international shrimp trade both to U.S. consumers and Florida processors. Imports of shrimp totaled 191.3 million pounds or 42 percent of total U.S. consumption in 1971. This was two million pounds less than the 193.3 million pound five year average of 1966-1970.

B. Problem:

A growing processing sector accompanied by a declining raw product sector causes the Florida shrimp processing industry to become increasingly dependent upon non-Florida shrimp for processing. The impact of past and potential changes in supply sources on market structure and ultimately on the competitive position of the Florida shrimp processing industry is uncertain and of concern to the industry and to the State because of the importance of the industry.

C. Hypotheses:

1. If the trend toward expanded imports of raw shrimp for processing continues, then the growth potential of the industry and the economic feasibility of locating shrimp processing firms in the State is reduced.
2. Intensified competition for and changes in the source of raw product supply create structural changes in shrimp processing (size, number, and vertical and horizontal integration of firms) and changes in market coordination and performance (purchase and sale practices, channels and margins) which place pressure on small firms to exit from the industry. Related hypotheses are that:

    a. As the age of firms in the industry increases, total sales increase thus providing a competitive advantage for these firms.
    b. Large firms tend to be more involved than small firms in research, innovation and product promotion thus providing a competitive advantage for these firms.

D. Objectives:

1. Develop basic information to describe the present characteristics (size, employment, volume, value, etc.) of the firms in the Florida shrimp processing industry.
2. Describe raw product supply, marketing channels and market structure in the Florida shrimp processing industry by studying the product flow and principal components of the Florida shrimp procurement and marketing system.
3. Delineate the organization, market behavior, and market performance of firms in the Florida shrimp processing industry by studying entry and exit, market concentration and vertical integration.
4. Determine the extent to which firms are involved in research innovation and product promotion.
5. Identify emerging changes and trends in procurement, processing activities, and product markets which will influence the shrimp processing industry in Florida.
6. Provide information for firms within the shrimp processing industry about the feasibility for expansion given recent and expected changes in product procurement and market structure.

V. Procedures:

The market structure approach to research provides economists with a basic theoretical framework for studying the behavior of enterprises. Empirical research involving market structure according to W.F. Mueller[8], may be divided into three broad categories. The first is measurement of the various structural variables which economic theory suggests are relevant to the performance; second, empirical determination of the economic bases of these variables; and third, specification and measurement of the interrelationships between market structure, firm conduct and industrial performance.

Secondary data will not be sufficient to satisfy all the objectives of this project so primary data will be collected in interviews obtained from

---

[8]Mueller, Willard F. "Empirical Measurements in Market Structure Research", Journal of Farm Economics. Vol. 43, December 1961, p. 371.

Florida shrimp processors. An inventory of shrimp processing firms must be developed to determine the population to be interviewed.

A. Marketing Channels:

Specific information to be collected for industry description will include number and location of plants, type of ownership (if multi-unit or single-unit companies), organization of the procurement and marketing system, organization of production based on type of shrimp and other products processed, and plant size by plant capacity, employment and 1972 production. The marketing system will be summarized by developing a product flow diagram of marketing channels which will describe the organization of the industry's supply, processing and distribution systems.

These marketing channels will encompass the supply of fresh shrimp (Florida and other U.S.) and of frozen shrimp (U.S. and foreign), market activities carried on by the processing plants, and final market outlets.

Raw product procurement will be analyzed in terms of access of each firm to raw supply, the different purchase channels used, prices paid and the transportation employed to bring the raw supply to the plant. Processing activities and final product destination will be described.

B. Evaluating Market Structure:

Market structure will be analyzed by studying the conditions of entry, market concentration, and market coordination. These major characteristics exert an influence on the nature of competition within the industry.

<u>The conditions of entry</u>. Entry and exit activities during the 1961-71 period will be analyzed by using secondary data available from the Florida State Chamber of Commerce. Entry related to age of firms will be studied by testing the possibility of a relationship between total sales and the number of years in the industry. Finally, processor responses will provide the information for discussing present entry conditions.

Entry and exit activities over the 1961-71 period will be analyzed by developing a plant entry activity ratio, plant stability ratio and plant exit activity ratio. Exit related to age of firms will be tested using ordinary least squares methods. Processor responses will provide information relative to entry conditions.

Market Concentration. Market concentration will be determined with data obtained from each firm's statement of annual volume and value for each type of product processed. The procedures employed will be the Lorenz Curve and the Gini Coefficient to help analyze industry concentration ratios developed for the entire industry and each type of processed product.

Conditions of scarcity and control of Florida raw shrimp stimulate the need to study the number of sellers as buyers of raw products because this situation may affect the structure of the industry. To determine the nature of this potential relationship, the percentage of firms controlling Florida's raw shrimp will be compared with overall concentration by product types to account for some differences in product line among firms.

Market Coordination. Degree of vertical integration prevailing within the industry will be determined by the number of related functions each firm performs. Functions may range from owning boats or unloading houses, operating breading plants, making boxes, owning transportation facilities and owning retail outlets to owning preprocessing facilities outside of the U.S.

Degree of horizontal integration will be established by determining the extent to which other seafoods were processed by the firms and the existence of processing facilities outside of Florida or the U.S. Reasons for horizontal integration will be explored.

C. Evaluating Market Conduct:

Three measures of market conduct will be considered. These are pricing behavior, product strategy and product promotion.

Pricing behavior. Price determination by processors for raw shrimp in the fresh domestic market and in the frozen market will be studied. The possibility of price leadership will be explored and compared with concentration and backward integration into the ownership or control of raw supply.

Pricing methods for finished products by firms in the industry will be analyzed. Pricing behavior will be compared with size of firms. Analysis of inventories will be related to price variations.

Further analysis will examine the relationship between price per unit and the number of units sold by each firm. Ordinary least square analysis will be used to investigate the hypothesis that price differentials prevail among firms in the Florida shrimp processing industry. The hypotheses to be tested will be that the price remains constant at all levels of quantity sold or that the price increases or decreases as quantity sold increases.

Product strategy and innovation. Product variation over time in both quality and design will be analyzed to determine product policy. Product variation will be examined in relation to size of firms and type of firms. Data will be collected on the effectiveness of large firms as compared to small in creating and introducing technological innovations. The chi-square statistical test will be used to determine if research and innovation are independent of firm size. Processor responses will be used to list product policy activities of the firms and to make comparisons with type of firms.

Product Promotion. Market structure theory suggests that fewness of sellers produces more incentives to advertise than large number of sellers. Since the Florida shrimp processing industry is composed of few sellers, this principle will be investigated using processor responses concerning advertising activities. Advertising will be compared with size of firms by means of the chi-square statistical test to determine if large firms are more likely to advertise than small firms in the industry.

D. Some Measures of Market Performance:

Margin analysis. Margin analysis for the industry and for individual firms within the industry will be accomplished to determine market performance. Gross margins will be analyzed by using ordinary least squares methods to regress total revenue and total expenditures on the quantity of raw shrimp bought. The resulting equations then will be compared with the average revenue and expenditure functions for the industry in order to explain possible differences in margins.

Margins also will be related to size of firms by means of the chi-square statistical test.

Productivity considerations. An attempt will be made to measure productivity as a means to evaluate performance. The percentage of plant capacity utilization during 1972 can be computed for the industry and for firms and for each group of firms with the same product lines will be computed.

E. Possible Industry Changes

Specific issues to be investigated in analyzing emerging changes and problems that may affect the industry include changes in raw product supply sources, changes in transportation and customers, problems with the supply of labor, utilization of plant capacity, and growth of firms over time.

VI. Budget:

Salaries	$6,220.00
Expendable equipment & supplies	150.00
Travel	1,000.00
Publication costs	150.00
Computer	200.00
Telephone	150.00
TOTAL	$7,870.00
Time available	One Year

VII. Published materials resulting from the project include:

Alvarez, Jose. "The Florida Shrimp Processing Industry: Economic Structure and Marketing Channels," unpublished M.S. Thesis, University of Florida, 1974.

Prochaska, Fred J. and Chris 0. Andrew. "Shrimp Processing in the Southeast: Supply Problems and Structural Change," Southern Journal of Agricultural Economics, Vol. 6, No. 1, July 1974.

Alvarez, Jose, Chris 0. Andrew and Fred J. Prochaska. "Dual Structural Equilibrium in the Florida Shrimp Processing Industry." Fishery Bulletin, Nacional Oceanic and Atmospheric Administration, Vol. 74, No. 4, 1977, p. 879-885.

Alvarez, Jose, Chris 0. Andrew, and Fred J.Prochaska. "Economic Structure of the Florida Shrimp Processing Industry," Florida Sea Grant Research Report, No. 9. 1975.

Andrew, Chris 0., Fred J. Prochaska and Jose Alvarez. "Florida Shrimp: From the Sea Through the Market," Marine Advisory Program, Florida Cooperative Extension Service, SUSF-SG-75-005, May 1975.

Case Two
MACRO ECONOMICS AND FINANCE

I. Project Title: The Brazilian Debt -- Effects of Real Interest Rates and Internal Factors on the Debt

II. Project Leader: Linda LaForest, Political Scientist and Agricultural Economist

III. Anticipated Duration: July 1, 1991 to December 31, 1991

IV. Problem Statement:

A. Problematic Situation

The research explores the causes of the current Brazilian debt situation. Between 1970 and 1985, Brazil's debt increased from $3.4 billion to $97 billion U.S., its interest and principal payments rose from $835 million to $10.8 billion U.S., and total debt as a percentage of GNP escalated from 8 to 37.6%. Critics point to the Brazilian government's penchant for foreign-financed spending as the cause of the growing indebtedness, but there has been little quantitative analysis of the world interest rate impact on the mushrooming debt. Our goal is to analyze Brazilian debt in terms of the external factor of world real interest rates and internal factors representative of Brazilian fiscal policy.

During the period from 1955 to 1970, Brazil followed a policy oriented toward vertical import substitution industrialization (ISI). The goal was to augment the previous period, which had emphasized horizontal ISI, by broadening the range of local production to include consumer durables, capital and intermediate goods. This was an attempt to reduce the drain on the balance of payments in the long run by reducing the need to import expensive machinery involved in domestic production (under horizontal ISI). However, this initially involved the importation of the capital-producing machines. The policy, implemented in 1956, was one of rapid industrialization based on full participation by foreign private investors. Transnational corporations (TNCs) were recruited to supply this capital.

The emphasis was switched to promoting diversified exports around 1970. The focus was on increasing the diversification of exports rather than only increasing the volume of a few exports. Between 1965 and 1972, the number of manufactured goods exported more than doubled. Attempts at vertical ISI also continued during this period. Increasing diversification of the sources of foreign investment was also a characteristic of this period, with more and more emphasis on private loans and less on grants or other government aid.

In 1973 there was global recession and oil prices quadrupled. The U.S. abandonment of the gold standard had the effect of devaluing the dollar and decreasing U.S. debt. In one author's opinion, the U.S. effectively "exported its debt." (7) These occurrences were accompanied by the growth in international banking (due in part to tighter U.S. regulation driving banks to expand internationally, i.e. out of domestic control) and in conjunction with the increasing availability and amount of petro-dollars. The Brazilian government sought to maintain its high levels of growth during this period by borrowing. They also attempted to undertake some massive projects aimed at infrastructure development, and these were financed mainly by foreign loans. Total debt between 1970 and 1986 increased from $3.4 billion to $97 billion U.S. The results were a large amount of debt per capita, a rapid increase in the rate of debt accumulation, and an increasing debt burden.

Oil price shocks were felt again in 1979 and 1980. In 1981 and 1982 world commodity prices dropped by 35%, international trade decreased and interest rates rose. The latter factor was of particular importance since most of the debt was held by private banks with floating interest rates. For example, one year during this period Citicorp increased its profits from Brazilian loans by 46%. Interest from Brazilian loans now generates 20% of Citicorp's worldwide profits. (7) In the 1960s, the IMF, with an expanding role began lending larger sums and setting conditions for economic stabilization, which were in turn taken as guidelines by the private lenders.

The resulting situation is one where Brazil relied on foreign currency to pay for imports and retire foreign debt; but due to lower export prices and scarcity of foreign loans it is difficult to develop the economy to the point where the quantity and/or quality of exports might be increased. Manuel Pastor, Jr. studied the effects of various internal and external factors on Latin American debt, and conclude that both contribute significantly to Latin American debt. (5)

B. Problem:

From the Brazilian perspective, the debt crisis means a siphoning of scarce resources away from social services and a lower standard of living. From the U.S. export market viewpoint, the IMF's emphasis on reducing imports and increasing exports means that not only can Brazil import less from the United States, but that more Brazilian exports will be in competition with U.S. products. The crisis affects the international, U.S. sponsored lending institutions on a more theoretical level; the relationship between the implementation of austerity measures and/or better conditions for foreign loans often conflicts with stated beliefs in democracy, basic human rights, and also global environmental concerns.

C. Hypotheses:

During the 1970s, the world recession and lower interest rates were conducive to borrowing by Brazil. In the 1980s the developed economies made an effort to "bleed" inflation from their economies. This was done by switching from a policy aimed at controlling interest rates to one targeted at controlling the growth in money supply. The switch to the latter policy resulted in increased real interest rates but controlled inflation. The significant increase in real interest rates meant that the "cheap" money previously borrowed by the debtor nations had to be paid back with more expensive dollars.

It is hypothesized that the foreign debt is a function of real world interest rates, the components of GNP (government spending, investment, net exports, and consumption) and the terms of trade. The level of foreign debt is obviously a function of borrowing by Brazil, but we would suggest also that it is significantly related to world interest rates. The purpose for the borrowed money can be deduced by looking at Brazil's government spending, net exports, investment, and consumption (national account).

A sorting out of these relationships will enable private banks, U.S. government policy makers, and international lending institutions to better anticipate the net results of their policies with regard to Brazilian loans. It is hoped that a deeper, economically holistic understanding of the forces driving the debt crisis can be used to solve and prevent (or at least mitigate) such crises in the future and lead to a more stable world economy.

D. Objectives:

The objectives are thus to disentangle and evaluate the factors contributing to the growth of the Brazilian debt. The hope is that such insights gained might guide private banks or international lending institutions when evaluating lending policies or implementing economic stabilization programs.

In particular, we wish to test the strength of the following relationships:

1. the effect of increasing world interest rates on external debt,
2. the effect of changes in government spending on external debt,
3. the effect of changes in investment on external debt,
4. the effect of changes in net exports on external debt,
5. the effect of changes in consumption on external debt, and
6. the effect of changes in the terms of trade on external debt.

V. Procedures:

The methods used will include the use of vector auto regression to evaluate the effects of the various macroeconomic variables on the growth and magnitude of Brazil's foreign debt. The relationship tested will be:

$$D = f(R, C, I, G, X, TT)$$

including lagged values of the variables and where Y = C + I + G + X and the values for C (consumption), I (investment), G (government spending), X (net exports) and Y (GNP) are taken from reference (2), D (debt) from reference (1), TT meaning terms of trade taken from the ECLA Statistical Yearbook of Latin America, and R (real interest rate) is taken from the rate of the 3 month U.S. Treasury bill and adjusted for inflation. For each of these measurements we will use quarterly data covering the period from 1970 to 1990.

Vector autoregression (VAR) is used because it avoids imposing many of the restrictions that are often applied to keep a model simple, therefore allowing a more "realistic" model to emerge. Each variable experiences lags, and this method is especially useful in examining and understanding dynamic systems. Using VAR one must first identify the model, variables involved, and length of time lags. Next one estimates the parameters, and then analyzes the results through the use of several tests. (4)

The initial process involves determining the lag length through the use of Sims' likelihood ratio test, testing the priors using the tests developed by Doan, et al., and testing for white noise, i.e. seeing whether any of the residuals in one time period are related to the residuals from other time periods. The second set of processes are testing the variables for Granger causality, testing impulse response functions, and examining the decomposition of variance and decomposition of forecast error. The Granger causality is simply an indication of the predictability of X on Y. The impulse response functions look at the mean path of the residuals, and trace the effect of a shock on a variable through subsequent time periods. The historical decomposition of forecast error and variance look at the amount of error due to the individual variables. Decomposing forecast error, one projects the growth of the debt starting from the initial conditions for one specific period. Then information on the actual observations of a particular variable are added to the forecast. The difference between the two projections is then attributed to the effect of the variable on which the information was added. Decomposing the variance is similar, except that instead of studying specific chunks of history the differences are analyzed between forecasts over differing numbers of lags. This second set of tests is merely a way of looking at different manifestations of the system, and from these observations the researcher then attempts to fit together the pieces into a coherent whole. (see reference (3) for more information)

Using this method, it is also possible to test for the effect of other variables (such as commodity prices) after running initial regressions. This allows more flexibility in designing the model.

VI. Budget:

The main cost in this project will be researcher time. The information is published in printed form, free from the library. The faculty advisory time is estimated to be 30 hours at $50.00 per hour. The time of the research assistant is estimated to be 320 hours at $15.00 per hour. Computer time and supplies is estimated to be equivalent to $300.00. Publication costs are estimated to be $200.00. Total cost therefore is expected to be $6800.00.

VII. Sources:

Data:

(1) World Debt Tables: External Debt of Developing Countries, Vol. II Country Tables, various editions, The World Bank, Washington, D.C. various years.
(2) International Financial Statistics, various editions, International Monetary Fund.

Method:

(3) Moss, Charles B., J.S. Shonkwiler and John E. Reynolds, "Government Payments to Farmers and Real Agricultural Asset Values in the 1980s," Southern Journal of Agricultural Economics, December 1989, pp 139-154.
(4) Jansen, Heinz, "Jamaica's Economic Decline: The Role of the Bauxite/Alumina Sector in a Time of High Energy Costs," Masters thesis, University of Florida, College of Agriculture, Food and Resource Economics Department, 1990.
(5) Pastor, Manuel Jr., "Current Account Deficits and Debt Accumulation in Latin America," Journal of Development Economics, Vol. 31, No. 1, July 1989, pp 77-97.

Background:

(6) Gereffi, Gary and Peter Evans, "Transnational Corporations, Dependent Development, and State Policy in the Semiperiphery," Latin American Research Review, Vol. XVI, No. 3, 1981, pp 31-64.
(7) Walton, John, "Debt, Protest, and the State in Latin America," in Power and Popular Protest, edited by Susan Eckstein, University of California Press, Berkely, California, 1989.
(8) Neuhaus, Paulo, "Floating Interest Rates and Developing Country Debt," Finance and Development, December 1982, pp 37-38.
(9) Hartland-Thunberg, Penelope and Charles K. Ebinger, Banks, Petrodollars and Sovereign Debtors, Lexington Books, Lexington, Massachusetts, 1986.

Case Three
PRODUCTION AND MANAGEMENT

I. Project Title: Improvement of Crop-Livestock Production Systems in Nandy Division of Wajir District in Kenya

II. Project Leader: William O. Odenya, Animal Scientist

III. Anticipated Duration: Three years

IV. Problem Statement:

A. Problematic Situation:

I. Characterization of the Area

The area receives about 40 inches of rainfall per year with 60% between April and June and 40% between October and January. Otherwise, drought persists during the remaining months. The area is hilly and rocky with soils ranging from black cotton soils to volcanic soils. The altitude is about 7,000 feet above sea level. The vegetation consists mainly of *Cynodon dactylon* grass species.

Tick borne diseases are a problem to livestock production, however, dipping facilities exist.

II. Characterization of the Farms:

The land holding per household is about 5 hectares with about 4 dependents. Farmers hold title to the land. Crop production, mainly consisting of maize, is alloted 3 ha. A few farmers intercrop maize with beans. When maize is near to harvest it is intercropped with cotton which is harvested at the end of the dry season before plowing starts in December. About 1,000 kg. of maize is harvested of which 70% is sold. Eight hundred kg. of beans is harvested of which 50% is sold. The remaining 30% of maize and 50% of beans is consumed within the household. Cotton is sold.

Cattle rearing consisting of 4 Zebu cows is alloted 2 ha. of land. Grazing is not improved. Up to 1965, the farmers in this area also raised sheep. An outbreak of Nairobi sheep disease killed nearly all the sheep in this area in 1965. After this incident, the area was under quarantine for five years. In 1975 a vaccine against the disease was developed and it no longer threatens sheep production.

During the dry season cattle feed on maize stover and other plant residues. Each animal produces about 1.8 liters of milk at the peak of lactation. About 50% of the milk is used for home consumption and the rest sold for cash. The cattle also provide manure for use in planting crops and also for fuel. Owning livestock is prestigious. There are also chickens in every household which supply meat and eggs for members of the family. They are also sold for cash.

Because most men have moved to the cities in search of wage employment, about 50% of the farms have females as the head of the household. Members of the family provide labour needed within the farm for crop and animal production. Females and children also perform duties including fetching water, looking for firewood and caring for animals.

B. Problem:

There is an insufficient amount of quality feed on a year round basis for cattle. Cattle also find it difficult to walk on rocky hills while grazing, thus limiting feed intake considerably. Food crop production takes priority over production of feed for cattle hence acreage for forage production cannot be expanded. The present milk production levels of about 1.8 liters per animal are too low to warrant intensive investment. Since milk is needed for household use, one alternative would be to invest more by purchasing higher yielding breeds, eg. Jersey. However, this is not possible since the area has tick borne diseases and the animals will also need more quality feed which is currently lacking. Given this situation, probably sheep are a better alternative to cattle. Sheep cannot provide milk but can provide meat as an alternative protein source. Compared to cattle sheep consume less forage per unit of output of meat. Sheep also have an advantage in efficiency of meat production because of greater prolificacy, earlier age of puberty, shorter gestation length and shorter growing period. Their smaller unit size makes them more adaptable to small farms and contributes to their ease of handling by women and children. They also utilize forage in hilly and rocky areas.

C. Hypotheses:

1. If quality forages and crop residues can be increased without decreasing food crop production, then milk production per head from the cows can be increased.

2. If excess feed available in seasons of excess, eg. during the rainy seasons, can be preserved and fed to animals during seasons of deficiency, then the animals can meet their nutrient requirements and produce normally at all seasons.
3. If some of the Zebu cattle can be replaced by sheep, better financial returns to the household can be realized.

D. Objectives:

1. In order to determine if forage production, quality and availability of crop residues may increase, research will:

   a. Evaluate cut and carry crops that can be planted along fence rows and wasted areas, and
   b. Determine some forage production interventions that can be applied to the cropping system without decreasing food production. For example, planting at higher densities to allow thinning and stripping.
2. Develop cheap ways of preservation of excess forage during seasons of excess.
3. Evaluate the potential of sheep within the production system, and:

   a. Evaluate complementarity that may exist between cattle, sheep and crops within the production system,
   b. Determine bioeconomic and socioeconomic constraints of incorporating sheep within the production system, and
   c. Obtain information on marketing and pricing of sheep and to ascertain if current input-output and risk factors favour incorporation of sheep within the production system.

V. Procedures:

1. To be able to accomplish objective number three, primary data will be collected through interviews with farmers in order to:

   a. Explore the social aspects of sheep production in the area,
   b. Determine literacy levels of household members,
   c. Establish baseline data on animal production,
   d. Develop an understanding of the relationship of cattle and crops, and determine the farmers' perception of sheep production, and
   e. Explore current management practices including animal health and disease control that are used by the farmers.

To be able to accomplish collection of primary data, four interviewers will be employed each for two weeks. The interviewers will be recruited through the extension service staff and will be from the research area. They will preferably be university students on vacation. Each interviewer will have training on data collection one week before the interviews are held. Each of them will be assigned ten farms.

2. Recruiting Farms

Data obtained from the interviews will be used to recruit the trial farms. To qualify for recruitment in the program, the household members must have shown some interest in the program. One of the household members must hold a high school diploma.

A total of 20 farms will be selected and divided into research areas. The interventions considered will be as below:

   a. Interventions to increase feed and crop residues from the existing crop system will include:

      i. Planting at higher than optimum densities to allow thinning and stripping.
      ii. Intercropping and relay cropping of forage crops in the food cropping system.

Maize will be planted in April and harvested in August at which time cotton will be planted. Cotton will be harvested in December and January. Cowpea for forage will then be planted to be ready by mid-March. Intercropping will be done by planting beans for grain in April which will be harvested in June-July upon which sorghum for forage will be planted and be ready by November. This will be followed by planting sunflower for forage which will be harvested by mid-March.

b. Interventions to increase feed and crop residues from existing non-cropped areas will include:

   i. Planting cut and carry forage grass; viz. Napier grass (*Pennisetun purpureun*). This will be planted along the contour on 1/4 hectares of land area. It will also help cut in control of soil erosion.
   ii. Planting grazing legumes (scattered) such as *Leucaena leucocephala* and non-traditional cut and carry crops such as pigeon pea.

Planting of Napier grass will be in April, while that of leucaena and pigeon pea will be in May.

3. Design of the trials will be as follows:

   a. Present traditional production system consisting of:

      i. Maize, beans and cotton on 3 hectares.
      ii. Four Zebu cows on 2 hectares.

   b. Present traditional production system with sheep:

      i. Maize, beans and cotton on 3 hectares.
      ii. Two Zebu cows and 10 Masai ewes on 2 hectares.

   c. Traditional production system with interventions:

      i. Maize, beans and cotton on 3 hectares.
      ii. Four Zebu cows on 2 hectares.
      iii. Interventions

d. Traditional production system with interventions and sheep:

   i. Maize, beans and cotton on 3 hectares.
   ii. Two Zebu cows and 10 Masai ewes on 2 hectares.
   iii. Interventions

Each trial will be replicated in 5 farms which will not be more than 2 miles apart.

4. Design feed preservation techniques for feed available in excess during seasons of abundance. The preserved feed is to be used during seasons of scarcity. Simple ensiling techniques using plastic bags and drums will be used. Use of hay as a roughage will also be encouraged. To be able to make high quality silage and hay, the forage plants including sorghum and the indigenous grass species will be cut at bloom stage.

5. Breeding

a. Sheep:

A breeding system of 3 lambings in 2 years is planned. In the first year the farmers will purchase pregnant yearling ewes with known parentage of twinning. The ewes will be allowed to rest for one month after lambing and then exposed to a ram. The rams will be leased from the Wajir Farmers' Training centre. At a year of age, selection will be done. All the males will be sold. The best 50% of the female yearlings will be retained as replacements. The replacements will be selected based on litter size and yearling weight.

b. Cattle:

Because of limitations of feed, the calving interval will be 22 months as shown. Since the number of females per farm will be insufficient to justify keeping a male for breeding, the farmers will lease bulls from the bull camps. Selection of females will be based on regularity of calving by the dam and on the heifer's own yearling weight.

6. Records to be Collected:

In each of the trial farms, the following data will be recorded by one of the family household members.

a. To meet the requirements of objective number one, the following data will be recorded:

   i. Representative samples of crop residue and forages fed to the animals will be collected at harvesting time. The samples will be taken to the National Agricultural Laboratories to be analyzed for crude protein, total digestible nutrients and ash. The analysis is done without any costs charged.
   ii. Visual appraisal of intake of crop residues and forages will be done by myself.
   iii. Yield in forages and crop residues.

b. For objective number two, the cost of drums and, plastic bags and strings for silage and hay making will be determined.

c. The following records will be needed for objective number three.

   i. Crops: Amount of each crop consumed, amount of each crop sold, and sale price of each crop.
   ii. Inputs: Labour -- household and hired labour is used in each of the crops.
   iii. Animals: Date of service, date of calving or lambing, and number of calves/lambs born and number that died.
   iv. In cows: Lactation milk yield, lactation length, amount of milk produced per day, amount of milk consumed by household per day, amount of milk sold and price sold for, and number of calves sold and price sold for.
   v. In sheep: Weight at one year of age.

vi. Expenditure: Amount of labour used in each livestock activity, and deworming drugs and dipping chemicals.

7. Labour Utilization

Labour utilization in farm activities within one year will be specified. The months of April, May and June are the peak periods for which labour will be required. May and June are especially critical since the children are in school. Hired labour is therefore required. About eight man-hours per day of hired labour is needed during this period.

8. Analysis of Data

For each of the trial farms, a balance sheet for revenue and cost will be worked out after every 12 months. Profit will then be determined. Labour use in every activity of the trials will be evaluated to assess viability of proposed innovations. The farmers' observations, perspectives and their tendency to adopt the changes will be evaluated by means of interviews. This will be done periodically and subjectively after every six months. Every farmer involved in the trial will be given an opportunity to see what other farmers (in different treatments) are doing. They will see one treatment at a time. After seeing the farms in other treatments they will be given a questionnaire where they will be asked to give points to the treatments. The treatment in which the farmer is, will be regarded as control. Dunnet's test will be used to compare control mean to the other treatment means.

At the end of the second year both gross and net income will be analyzed. Since the design is completely randomized, the model will be as follows:

$$Y_{ij} = N + a_i + e_{ij}$$

in which:

Y_{ij} = j^{th} observation on the ith treatment ($i=1,\ldots,4$)

N = overall mean of all observations

a_j = A random error associated with the jth treatment

e_{ij} = A random error associated with the jth observation on the ith treatment

Assumptions:

i. The a_j are a random sample of treatments from a population which is normally distributed, has a mean of zero, and has a common variance, ^2a

ii. The e_{ij} are randomly distributed with zero mean and variance equal

The analysis will give an indication whether gross income/net income differ among treatments. If the differences in treatment means are found significant, orthogonal contrasts will be done.

For example:

i. To compare treatments with sheep with all others without sheep.

ii. To compare treatments with interventions with those without.

iii. To compare the differences between treatments with sheep and those without sheep in the absence and presence of interventions.

VI. Budget

Transport	$1,600
Subsistence allowance	3,840
Salaries of interviewers	800
Equipment and supplies	500
Incidental expenses	1,000
	$7,740

Time available 3 years

Case Four
RESOURCE MANAGEMENT

I. Project Title: Economic effects of imposing harvest restrictions on the Atlantic swordfish fishery.

II. Project Leader: Lee Bouchelle, Marine Economist

III. Project Duration: January 1, 1991 through December 31, 1991

IV. Problem Statement:

A. Problematic Situation:

The North Atlantic swordfish (*Xiphius gladius*) has recently been the subject of an increasing number of proposed domestic and international regulatory measures. These measures are designed to reduce current fishing mortality which is believed to place the swordfish stock in danger of recruitment failure; the reduction in biomass weight of fish entering the catchable population during a specific time period. A proposed amendment to the existing fishery management plan (FMP), as well as an international agreement to limit swordfish harvest, will contribute to the formulation of management measures that are intended to rebuild the spawning stock to a level that will reduce the likelihood of recruitment failure. These new management initiatives are the result of an evaluation of the swordfish stocks by scientific panels, international negotiations, input from public hearings and economic analysis. At the current time the National Marine Fisheries Service (NMFS), with direction from the Secretary of Commerce, is responsible for the formulation and implementation of such management measures.

B. Problem:

Fishery biologists have determined the North Atlantic swordfish fishery to be overfished. Management concern over the health of swordfish stocks was prompted by the fact that the total number of swordfish caught by domestic fishermen has increased substantially from 1978-1989 (Figure 1), while the average weight per fish landed has declined steadily over the same time period (Figure 2). Analysis of the 1987 stock assessment by fishery biologists indicates that the adult spawning stock biomass (SSB) in 1987 was about 40 per cent of the 1978 level, while the mean size of harvested fish has declined continuously from 115 pounds to 60 pounds dressed weight (dr. wt.). A 60 pound fish is approximately 2-3 years of age but not sexually mature. If these trends continue, domestic yield is projected to fall from 10.3 million pounds in 1989 to 4.95 million pounds (dr. wt.) in 1991. If the predicted downward trend in landings occurs, only 0.11 million pounds (dr. wt.) would be harvested during the year 2000. Even taking into account higher prices offered due to decreased supplies, projected revenues would be $.44 million (South Atlantic Fishery Management Council, 1990). These reductions in total revenues do not incorporate other negative economic factors such as unemployment in the production and processing sectors, and reductions in domestic welfare.

FIGURE 1 Reported U.S. landings
from mgmt. unit (total # caught)

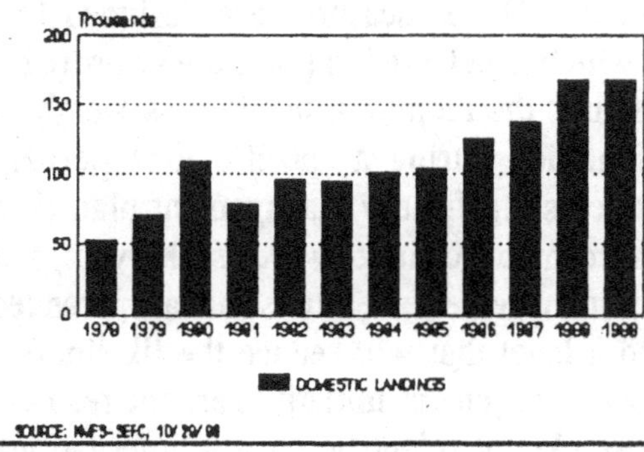

Management measures that have been suggested to combat this problem are a variety of harvest restrictions that may have significant economic impacts. The options currently being considered by NMFS include those proposed by the International Commission for the Conservation of Atlantic Tuna (ICCAT) and the South Atlantic Fishery Management Council (SAFMC) or some combination of elements contained in both proposals. The ICCAT recommendations include a prohibition on the taking of swordfish under 25 kilograms (approximately

55 pounds whole weight or 41 pounds dr. wt.) with a 15 per cent tolerance on the taking of fish less than 25 kilograms, and a 15 per cent reduction of fishing mortality on fish greater than 25 kilograms (NMFS, 1990). The SAFMC developed an amendment to the current

FIGURE 2 Reported U.S. landings from mgmt. unit (mean lbs dr wt)

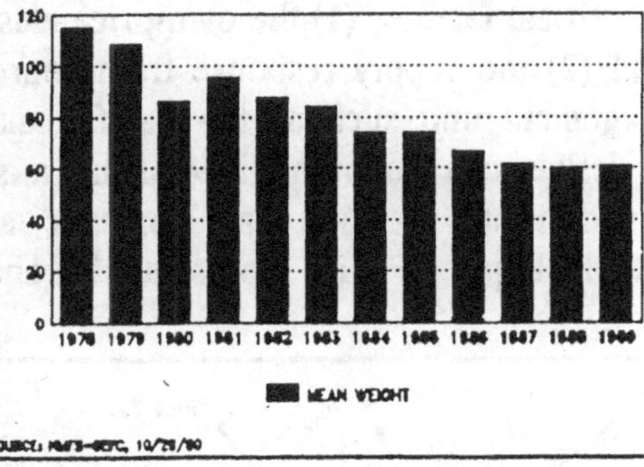

FMP which would reduce domestic landings and imports of swordfish from the management unit by 31 per cent. Regardless of which option the Secretary of Commerce ultimately approves, some form of harvest reduction will be required from both domestic and international fishermen targeting swordfish from the North Atlantic stock. Hence, swordfish supplied to domestic markets from North Atlantic sources will be reduced with potentially significant economic effects. Short term costs such as loss of domestic surplus and the loss of revenues in countries that export swordfish to the U.S. are being weighed against the long term benefits of reducing the current harvest of swordfish: thereby ensuring the survival and growth of the North Atlantic swordfish stock.

C. Hypotheses:

1. If domestic and import quotas are implemented thereby reducing the supply of swordfish available in the market, then there will be a short term price increase and a consequent loss in consumer and producer surplus.

2. If imports from other stocks such as the Pacific respond to the shortfall in supply from the Atlantic stocks, the price effect may be minimized.

D. Conceptual & Theoretical Framework

Reductions in swordfish supplied to the market may result in changes in earnings to domestic and international fishermen targeting North Atlantic swordfish stocks. The reductions in North Atlantic swordfish supplies may also result in changes in prices paid by consumers. The actual magnitude of these changes in producer and consumer welfare will depend upon two critical factors, (1) the own-price elasticity of demand for swordfish and (2) the supply response from sources of swordfish outside the management unit such as the Pacific and Mediterranean regions. These issues can be highlighted using a simple graphical representation of domestic demand and domestic and international sources of swordfish supply. Figure 3 presents this analysis.

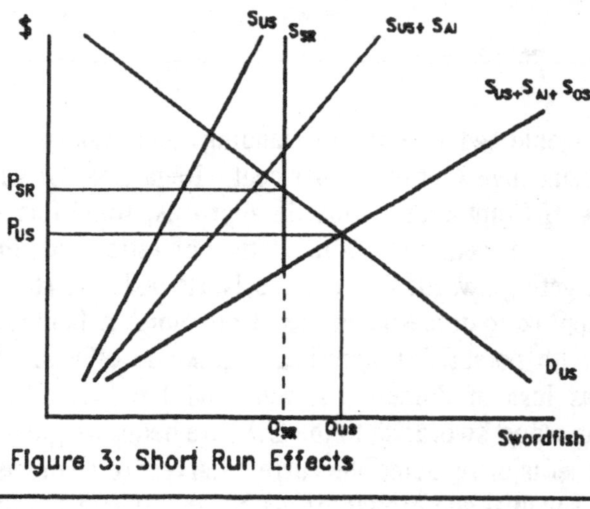

Figure 3: Short Run Effects

The U.S. demand for swordfish is represented in Figure 3 as D_{US}. The quantity of swordfish supplied to the domestic market consists of the horizontal summation of the individual supply curves from domestic and international sources. These are shown as S_{US}, S_{US} + S_{AI}, and S_{US} + S_{AI} + S_{OS}, for domestic, domestic plus imports from North Atlantic sources, and domestic plus North Atlantic imports plus Other Sources (Pacific and Mediterranean imports) of swordfish, respectively. At the initiation of the harvest reduction measures, a greater than 15 per cent reduction in total swordfish supplied from North Atlantic stocks would result due to the combined effect of the 15 per cent reduction on larger fish and the prohibition on harvest of fish less than 25 kilograms. This reduction would cause the total supply curve to become vertical (S_{SR}) at that point where swordfish supplies are below current levels *in the short run*. Short run is defined as the time period that the producer is "constrained to use the equipment in place, that is, unwilling to adjust fully and immediately to a new scale of operation" (McCloskey 1988). At this level of supply, the domestic swordfish price rises from P_{US} to P_{SR}. Note, however, that the aggregate supply curve may also shift upward due to increased fishing costs because smaller size fish can no longer be harvested. This possibility is not shown, in order to simplify the analysis. Consumers become worse off due to the higher price. However, the actual change in price will depend upon the own-price elasticity of demand with prices going up less the more price elastic is the demand curve. Swordfish producers may be made better or worse off depending upon whether the proportional change in price is more than (demand is inelastic) or less than (demand is elastic) offset by quantities supplied.

The changes of consumer and producer welfare just described are short run phenomena. In the long run, assuming Atlantic producers are bound by harvest restrictions, the changes in consumer and producer welfare will depend upon the ability and propensity for sources outside the management unit to increase swordfish quantities supplied to the U.S. market. Figure 4 provides an illustration of these possibilities.

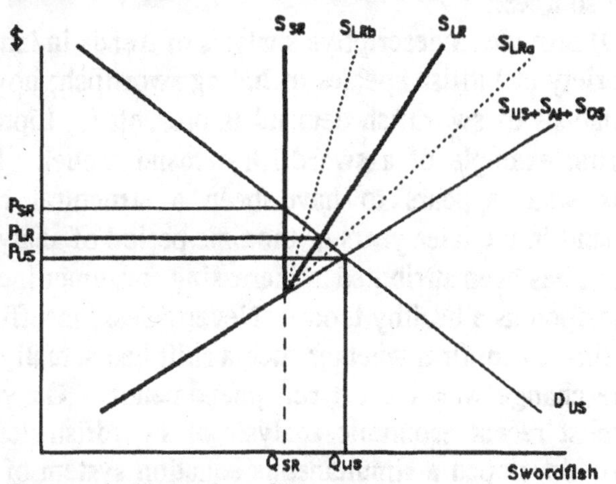

Figure 4: Long Run Effects

To simplify the illustration, Figure 4 shows only domestic demand and

scenario previously described is shown along with the long run scenario. In both cases North Atlantic supplies cannot respond to the higher price due to harvest restrictions. Therefore, the shape of the long run supply (S_{LR}) curve depends only on the supply response of imports from other sources. If Pacific and Mediterranean sources do not increase imports at all then the long run and short run supply curves are the same (S_{SR}). The more likely case is that fishermen will increase harvest of Pacific swordfish stocks to some degree. The long run supply curve S_{LR} shows one possibility of a partial response by these fishermen. The long run supply curves S_{LRa} and S_{LRb} show other possibilities where other stocks may respond respectively somewhat more and somewhat less than S_{LR}. As long as there is some long run Pacific supply response, reduced consumer welfare in the short run will be mitigated by lowered prices in the long run. The long run effect on North Atlantic domestic and international producer's well being will depend, once again, upon the elasticity of demand and the magnitude of the supply response from other sources.

In order to estimate the actual changes in consumer and producer welfare as a consequence of management restrictions on North Atlantic swordfish, the theoretical relationships illustrated in Figures 3 and 4 would need to be empirically estimated. Unfortunately, little empirical work has been done to estimate the US demand for swordfish, nor has adequate research been conducted to estimate supply functions for swordfish imported to the U.S. from North Atlantic, Pacific, and Mediterranean sources.

King (1989) provides a descriptive analysis of trends in landings and prices for a variety of billfish species including swordfish; however, no quantitative analysis of swordfish demand is presented. Lipton (1986) provides the first example of a swordfish demand model. However, Lipton reports what appears to have been a structural change in swordfish demand in the later years of the time period of analysis. This structural change has been attributed to increasing consumer incomes and the image of seafood as a healthy food. Nevertheless, insufficient data existed at that time to confirm whether such a shift had actually occurred or whether the change was a short run phenomenon. Gauvin (1990) provides the most recent economic analysis of swordfish demand and supply. Gauvin developed a simultaneous equation system of domestic demand and aggregate supply (domestic supply plus all imports). Gauvin found domestic demand to be slightly elastic (-1.04) to somewhat inelastic (-.74) (Gauvin, 1990).

No prior research has attempted to estimate supply responses from domestic and international sources. Gauvin (1990) did include a supply equation in the model but no distinction was made between domestic and import supplies and no distinction was drawn between Atlantic and other supplies. This latter deficiency is of particular importance because, as was shown earlier, Atlantic fishermen will be subject to harvest restrictions and will be unable to respond to increased swordfish prices which will not be the case for Pacific and Mediterranean fishermen. The supply response of Pacific fishermen is of particular importance if the economic effects of Atlantic swordfish management strategies are to be properly evaluated. Therefore, the proposed research has the following objectives.

E. Objectives:

1. Describe the domestic swordfish industry, its evolution, current user groups, gear type, etc.
2. Describe trends in world landings and the growing role of imports in the domestic market as well as the role that other markets (European) may play in swordfish trade.
3. Develop a model that can be used to estimate supply and demand and to measure the loss in domestic surplus. Specifically this can be done by empirically estimating the own-price demand elasticity as well as the output elasticities for domestic and import supplies of swordfish. Import supplies include swordfish from the Atlantic, Pacific and Mediterranean stocks. The own-price elasticity of demand can be used to predict short-run price changes once harvest restrictions have been imposed. From this predicted price, a supply response from the Pacific and Mediterranean stocks can be estimated as well as the subsequent losses in domestic surplus.

V. Procedures:

A. Description of domestic swordfish industry:

This will include an historical overview of the domestic fishery with information such as effort expended, areas fished, as well as technological evolution within the fishery. Gear type, in addition to numbers of vessels participating in the fishery is important in determining the amount of effort in the fishery. Logbook data will be utilized in compiling this information.

In addition, domestic production and consumption data will be surveyed from 1959-1990. The purpose of this is to determine patterns in production and/or consumption in the domestic swordfish industry.

B. Trends in world landings and imports:

Data fom the National Marine Fisheries Service (NMFS) and the United Nations' Food and Agricultural Organization (FAO) will be used to determine world landings and thus world supply of swordfish. Import data from the Department of Commerce will show the growing importance of imports as a component of U.S. supply. Imports will be categorized as imports from specific regions such as Atlantic, Pacific, and Mediterranean etc. The viability of the EEC as an alternative market for both domestic and international swordfishermen will be addressed. The availability of alternative export destinations is relevant to any management measures considered by domestic and international agencies.

C. Empirical estimation of world swordfish supply and domestic demand:

A simultaneous equation system will be developed to estimate the own-price elasticity of demand for swordfish. Simultaneous equation systems are commonly used to estimate supply and demand relationships for agricultural commodities (Wohlgenant, 1989; Willet and French, 1991) and have been used to describe market relationships for shrimp (Adams, Prochaska, and Spreen, 1987) and for salmon (Herrmann and Lin, 1988; Lin and Herrmann, 1988; Lin, 1988). Only Gauvin (1990) has tried to use a simultaneous equation system approach to estimate swordfish demand and supply.

The aggregate amount of swordfish demanded in the United States is expected to be negatively related to swordfish price, positively related to prices of substitute products, and positively related to personal income. Swordfish supplied to the U. S. market can come from any one of three sources: domestic production in the North Atlantic, imports from countries fishing from North Atlantic stocks, and international sources fishing from Pacific or Mediterranean waters.

The amount of swordfish that fishermen are willing to supply is expected to be positively related to the price of swordfish, negatively related to the price of alternative target species such as tuna, and negatively related to costs of production. Suppliers of imported swordfish are expected to respond to the same variables as that of U. S. fishermen. However, other international markets such as the European community and Japan are assumed to be alternative shipping destinations. Therefore, the simultaneous equation system consists of an U.S. aggregate demand equation, an equation for U.S. supply, and one equation each for import supply from North Atlantic stocks and from other stocks respectively. Finally, an identity equation is added to assure that markets clear. The system of equations is as follows.

(1) $Q^{D}_{US} = f(P^{US}_{SW}, P_{SUB}, Y)$

(2) $Q^{S}_{US} = f(P^{US}_{SW}, P^{US}_{T}, C_{US})$

(3) $Q^{S}_{AI} = f(P^{US}_{SW}, P^{EC}_{SW}, P^{J}_{SW}, C_{AI})$

(4) $Q^{S}_{OS} = f(P^{US}_{SW}, P^{EC}_{SW}, P^{J}_{SW}, C_{OS})$

(5) $Q^{D}_{US} = Q^{S}_{US} + Q^{S}_{AI} + Q^{S}_{OS}$

where: Q^{D}_{US} is aggregate demand for swordfish in the United States, Q^{S}_{US} is total swordfish supplied by domestic sources, Q^{S}_{AI} is total swordfish imported from North Atlantic stocks, Q^{S}_{OS} is total swordfish imported from stocks other than North Atlantic, P^{US}_{SW} is United States swordfish price, P_{SUB} is an index of swordfish substitute seafood price, P^{EC}_{SW} is an index of European Community swordfish price, P^{J}_{SW} is Japanese swordfish price, P^{US}_{T} is United States tuna price, C_{US} is an index of United States fishery production costs, C_{AI} is an index of North Atlantic importing country fishery production costs, and C_{OS} is an index of importing country fishery production costs from sources other than the North Atlantic.

The demand equation is similar to that used by Gauvin with the exception that an index of substitute seafood prices is considered to be a better proxy for swordfish substitutes than is beef as was used in by Gauvin (1990). The use of an index of seafood prices should reduce the collinearity problems encountered by Gauvin when only tuna was used as a swordfish substitute. The seasonality variable used by Gauvin is not included as the author found no evidence of seasonality in swordfish demand.

The supply equations also are similar to those which Gauvin estimated except that we treat import supplies separately from domestic supply. Accordingly, world market prices are included in the supply equations. Gauvin used an index of diesel fuel price as a proxy for vessel operating costs. This measure combined with wage rate indices will be used to measure the operating cost variable for domestic and international swordfish suppliers.

The simultaneous equation system described above has five equations and eight parameters to be estimated. By specifying functional forms, the entire system can be estimated using instrumental variable techniques (Judge et. al, 1985). From the parameter estimates, own-price and income elasticities of demand and supply response for each of the suppliers can be calculated. Additionally, welfare effects can be measured.

VI. Budget

A. Salaries	
Research Assistant (1/2 time M.S.)	13,467
B. Travel	
1. Southeast Fisheries Center, Miami	620
2. NMFS Southeast Regional Office, St. Pete	220
C. Miscellaneous Costs	
1. consumable supplies	200
2. telephone	50
D. Total Budget	14,557

VII. References:

Adams, C. M., Prochaska, F. J. and T. H. Spreen. "Price Determination in the U.S. Shrimp Market." Southern Journal of Agricultural Economics. 19(1987):103-112.

Gauvin, J. R. "Estimation of Supply and Demand Functions for Broadbill Swordfish in the United States." Amendment 1 to the Fishery Management Plan for Atlantic Swordfish. South Atlantic Fishery Management Council, Charleston, S.C. 1990.

Herrmann, M. and B. H. Lin. "The Demand and Supply of Norwegian Atlantic Salmon in the United States and the European Community." Canadian Journal of Agricultural Economics. 36(1988):459-471.

Judge, G. G., W. E. Griffiths, R. C. Hill, H. Lutkepohl, T. C. Lee. The Theory and Practice of Econometrics 2nd Ed. John Wiley and Sons, New York, New York. 1985.

King, D. M. "Economic Trends Affecting Commercial Billfish Fisheries." Planning the Future of Billfish. National Coalition for Marine Conservation. 1989.

Lin, B. H. and M. Herrmann. "An Econometric Analysis of Atlantic Salmon Markets in the U.S. and France." Alaska Sea Grant Report 88-5. 1988.

Lin, B. H. "The Demand for Atlantic Salmon in Canada: Issues of Functional Form and Parameter Stability." Alaska Sea Grant Report 88-6. 1988.

Lipton, D. W. "The Resurgence of the U.S. Swordfish Market." Marine Fisheries Review. 48(1986):24-27.

McCloskey D. N. The Applied Theory of Price 2nd Ed. Macmillan Publishing Company, New York, New York. 1985.

National Marine Fisheries Service. "Public Hearings on Management Options for Atlantic Swordfish Scheduled." News Bulletin. National Marine Fisheries Service, Southeast Region. 1990.

South Atlantic Fishery Management Council. Amendment 1 to the Fishery Management Plan for Atlantic Swordfish. South Atlantic Fishery Management Council, Charleston, S.C. 1990.

Willet, L. S. and B. C. French. "An Econometric Model of the U.S. Beekeeping Industry." American Journal of Agricultural Economics. 73(1991):40-54.

Wohlgenant, M. K. "Demand for Farm Output in a Complete System of Demand Functions." American Journal of Agricultural Economics. 71(1989):241-252.

Case Five
THEORY AND INTERNATIONAL COMPETITIVENESS

I. Project Title: Modelling International Competitiveness: A Multilateral Econometric Approach

II. Project Leader: Patrick Antoine

III. Project Duration: 18 months

IV. Problem Statement

A. Problematic Situation

The internationalization of the world economy, fostered by the free movement of capital, technology and traded goods, has resulted in countries becoming more interdependent. Trade as the engine of growth in virtually all economies has been enhanced partly by the General Agreement on Tariffs and Trade (GATT), which encourages trade liberalization and integrated markets for goods. Since virtually all economies have become integrated through trade, national policies as they relate to trade can have serious implications for the pattern of trade between countries. The process of internationalization and the attendant integration of capital markets has meant a more rapid transmission into the global economy of national disturbances resulting from domestic and trade policy distortions. Hence, product flows between trading partners will be influenced directly by national government policies, exchange rates, national income and domestic policies. It is in the context of international trade that competitiveness assumes greatest importance.

This study will examine international competitiveness in agriculture between the U.S. and Canada from 1961 to 1989. Warley (1977), argues that no sector better demonstrates the deepening interdependence of the world economy than agriculture. The implication of this 'deepening' as Warley calls it, for the U.S. and Canada has been increased competition for markets, owing to the similar nature of the agricultural commodities produced and exported by both countries. Warley argues that the strength of the North American economy, and its ability to influence world economic events is highly dependent on the performance of the U.S. and Canadian agricultural sectors in the world economy. Consequently, the issue of international competitiveness in agriculture between the U.S. and Canada seems particularly amenable to analysis via the theoretical model proposed in this study.

Achieving and maintaining competitiveness depends on a gamut of both demand and supply factors which will determine which commodity will be imported or exported. In the short-run however, a country's ability to compete will be determined by the ability to offer similar commodities at more competitive prices than other countries. In the case of similar products though, a country's demand would most likely be invariant to the source of supply. National government policies will also affect the competitiveness of a country's exports, along with the explanations of resource costs and resource endowments advanced by conventional trade theory. Empirical investigations of these factors as well as the forces that determine both absolute and comparative competitiveness are critical to the understanding of competitiveness in the short-run. Empirical measures of international competitiveness should therefore account for the influences of real factors (those determining productivity), monetary factors (wages, prices and exchange rates), and structural factors the most important of which are supply conditions, along with the effects of national government policy.

B. Conceptual Problem

Competitiveness may either be defined in terms of the composition of output or the composition of employment. Four general definitions have emerged in the literature of which two are related to the macroeconomy and two to the microeconomy. First, and perhaps the most common definition, is implicit in studies linking the trade balance to competitiveness, (Hicks, 1959; Johnson, 1968). This definition holds that a deteriorating trade balance is symptomatic of declining competitiveness. A second related definition is based on sectoral trade balances in specific industries or products and admits to the possibility that sectoral competitiveness may be declining congruent with an improving trade balance. By and large, these are macro-measures of competitiveness and the oft-cited rationale of macroeconomic policy targeted at reversing the downward trend in competitiveness, is apt to be unsuccessful unless accompanied by specific measures to eliminate the trade deficit.

It is to the remaining two definitions that attention must be turned, because of their role in microeconomic studies. The market-share definition of competitiveness assumes that, over time, an increase in market share is evidence of increasing competitiveness, either at the sector or industry level. Again the same criticism levied on the second definition rings true for this notion of competitiveness; declining market-share is often not a signal of declining sectoral trade balance. The final definition relates various notions of productivity to competitiveness. McCulloch (1986) argues that whereas causality between greater productivity gains at the industry level and enhanced competitiveness are automatically assumed, neither result is logically implied.

Competitiveness, in its most sterile sense, refers to the ability to win or achieve some goal. Balassa (1962), defined it as the ability to sell under conditions of fixed exchange rates. Subsequently, Islam (1983), contended that competitiveness could also be conceived in terms of changes in costs, prices, or exchange rates; so that a depreciation in one country's currency relative to the currency of another country could signify a gain in competitiveness for the country whose currency had depreciated and a loss in competitiveness for the country whose currency had appreciated in relative terms. The commonality between these two approaches is that they both define competitiveness in terms of the ability to capture market shares. In industry (e.g. agriculture, manufacturing), it usually refers to the ability to sell profitably both at home and abroad. It is to this latter definition of competitiveness originally advanced by Balassa and subsequently amended by Islam that this study subscribes.

The reason for interest in international competitiveness as opposed to competitiveness per se is best explained by Porter (1990), who points out that forces of competitiveness may be country specific and not firm specific as often believed. Still as Porter, and Ahearn et al. (1990) contend, there is no general theory of competitiveness because it is not strictly an economic concept. They, however, go on to argue that the principle of comparative advantage is a key element of competitiveness.

Measures of international competitiveness are often used as a performance criterion in analyzing sectoral or macroeconomic performance across countries. In addition to quantifiable variables, commonly used measures also encompass qualitative factors such as product specialization, product quality, and the value of after-sales service, among others. While the inability to account for these non-quantifiable though important factors is undoubtedly a drawback for some purposes, it may not be critical if changes in relative competitiveness, rather than absolute levels of competitiveness, are used in analyzing trade trends. To the extent that this is possible, some of the failure to explicitly account for non-quantifiable factors may be mitigated.

C. Problem Statement

Interest in international competitiveness in industrialized countries resurfaced in the 1970s amidst unprecedented threats to market shares from Japan and the newly industrialized nations of Taiwan and Korea. In the economic literature, the research agenda took considerably longer to react and theoretical work on international competitiveness did not begin appearing until the early 1980s. Thompson (1981), reviewed over 300 studies, most of which were in the areas of international trade and agricultural sector modelling, but which in some way related to the concept of competitiveness. Brinkman (1987), however contends that the various approaches utilized by these studies were inappropriate for assessing competitive performance between countries, since they only dealt with the competitiveness issue in an indirect manner, through domestic prices in the specification of supply and demand factors and through transport cost and policy variables. Only recently have attempts to directly assess international competitiveness appeared in the theoretical literature. In many cases though, the sole focus has been on measuring international competitiveness, and in this regard the work of Vollrath (1987) is noteworthy.

A subject of considerable importance and one in which theoretical progress has been slow, is the examination of the factors that determine or explain competitiveness between countries. This is to some extent understandable, since the lack of a universal definition for competitiveness makes any attempt to attribute competitiveness to certain specific factors elusive at best. Still, the ability to define international competitiveness as the result of a number of specific causal processes, some quantifiable, others less so, can further advance its understanding. It is still regarded as a 'black-box' by many economists, and remains an area of concern for economic entities that must compete in global markets.

Competitiveness is also widely believed to be the outcome of a number of dynamic forces acting in concert. This belief is partly supported by the view that the global economy is in a constant state of flux. Many believe that any attempt to develop a serious concept of competitiveness must of necessity account for changing international and domestic factors. As an industry loses competitiveness, economic agents must either shift to another economic endeavor or leave the industry altogether. Any measure of international competitiveness must account for these changes in the global marketplace.

There has been a proliferation of articles which allude to international competitiveness either directly or indirectly. While asserting competitiveness as a critical issue on the national agenda, they do not possess the theoretical rigor or the empirical content through which a determination of the patterns or magnitudes of trade between countries can be achieved.

In sharp contrast to the large number of studies which indirectly address international competitiveness (international or interregional trade models), studies which directly address competitiveness in agriculture are relatively few. Interest in developing an alternative framework for the measurement of international competitiveness, is underscored by the associated difficulties with empirical implementation of the principle of comparative advantage, in conjunction with slow theoretical progress in the development of measures of international competitiveness capable of explicitly accounting for the factors that determine or explain competitiveness between countries.

D. Hypotheses

Many factors influence competitiveness and several of them have already been mentioned in the introduction to the study. Regrettably, not all of them can be examined in a study of this nature. This study will focus on a number of specific factors, considered common to the theoretical literature on competitiveness.

The first proposition of the study is that a multilateral theoretical model of international competitiveness can be developed based on previous research that will:

(1) measure the impact of productivity, economies of scale, and technological differences between countries, as well as wage, price and exchange rate differentials, and national agricultural and trade policies instituted by individual governments on a measure of international competitiveness; and

(2) deduce the relationship between these factors and the relative importance of each to changes in the measure of international competitiveness.

The second major proposition is that an empirical model based on the economic theory of production can be specified which explicitly accounts for differences in the structure of production between countries thereby providing insights concerning their international competitiveness.

E. Objectives

It is the intent of this study to develop a general model of international competitiveness which will measure multilateral sectoral/industry performance between countries. This will be accomplished using a restricted profit function, in the presence of short-run dis-equilibrium and non-constant returns to scale.

In developing this alternative framework, it is the intent of this study to develop an empirically estimable model of international competitiveness which will facilitate analysis of both the absolute and relative importance of various determining factors on measured multilateral competitiveness performance. In addition to determining the magnitude and extent of the suggested linkage between these determining factors and international competitiveness, the study will also develop a framework through which these relationships can be discussed. While the model developed in this study is quite general, it will be applied to competitiveness in agriculture, if only because of the small number of studies undertaken thus far on this sector. The importance of this research for the agricultural sector is further accentuated by its timeliness, since the U.S.-Canada Free Trade negotiations have only recently been concluded.

The specific research objectives and goals of the empirical analysis are to:

(1) test the propositions arising out of the theoretical analysis with respect to the structure of aggregate agricultural production in Canada and the United States;
(2) compute empirical estimates of the model using data for the U.S. and Canada from 1961-1989;
(3) utilize the relevant parameters as the basis for inquiry proposed in goal (1) above, and the first hypothesis stated; and

(4) provide insights into the policy implications for the agricultural sectors in both the U.S. and Canada, based on the results of the model.

F. Theoretical Model

While no general consensus exists on the definition of competitiveness, considerably more agreement exists as to the factors which contribute to competitiveness. Morici (1984), posits that among structural factors, supply conditions play a principal role in determining a country's comparative advantage. He further argues, that the amount and quality of physical capital availability per worker, the quality and skill of the labor force, the stock of research and development capital per worker and the availability of natural resource endowment all contribute to overall productivity and hence to competitiveness. Dogramaci (1981), also argues that productivity is a key determinant of international competitiveness. Productivity advances reduce unit cost of production, increase output and reduce the price of products in competitive markets. In addition higher productivity levels allow factors to receive higher renumeration, while ensuring that the country retains its competitive edge.

Other factors which are likely to play a pivotal role in the determination of competitiveness are technical change and economies of scale. On the issue of technical progress, Harriss (1981) points out,

> "That scientific and technological advances have brought incalculable advantages in productivity. Many are striking and obvious, many indirect. Often, accident has played a part."
> (p. 43).

Numerous other studies, Shumway (1977), Ruttan (1980), Norton and Davis (1981) also assert the importance of technical change to productivity and hence to competitiveness. Economies of scale may also be an important determinant of international competitiveness in some industries. While the preponderance of evidence suggests an absence of scale economies in agriculture, evidence of economies of scale has been found in other studies, (Griliches, 1964). It therefore appears that the question of economies of scale in agriculture remains an open one.

The model is based on the assumption that economic agents will produce and export commodities based primarily on the profit motive.

Defining international competitiveness in this manner would not be inconsistent with the conventional models of comparative advantage but for two revisions. The first is that the profit function is specified in post-trade prices, and the second is that it incorporates distortions and deviations from the free trade assumption. The Measure of International Competitiveness (MIC), framework relaxes the assumption of identical technologies across countries maintained by the Heckscher-Ohlin-Vaneck (HOV) approach and the factor fixity assumption of the Ricardo-Viner (RV) approach. This allows differences in factor usage and efficiency to influence competitiveness over time. Adherence to the view that inter-country differences are central to trade, constitute ipso facto acceptance of resource endowments and technology explanations as key components in the determination of the basis for trade. The MIC model therefore makes it possible to examine the implications of both relative technology differentials and relative rates of technological change between countries.

Given this the MIC between country i, (base country) and country j (any other country) is defined as:

$$MIC_{ij}^{t} = \frac{\pi_{i,t}(P_{sit}, v_{jit}, z_{kit}, T_{it})}{\pi_{j,t}(P_{sjt}, v_{jjt}, z_{kjt}, T_{jt})} \qquad (1)$$

P_{sit} is the industry\sectoral product price in country i at time t; v_{jit} is a vector of variable inputs prices at time t in country i; z_{kit} is a vector of quasi-fixed or primary inputs at time t in country i; T_{it} is an index of disembodied technical change in country i at time t. Variables corresponding to country j are defined in a manner totally analogous to the definitions given for country i.

The ratio of restricted profits measured in the same currency between two countries producing homogeneous products with similar inputs, constitutes the basis for the measure of competitiveness, MIC[9]. Restricted profits may be thought of as a residual measure composed entirely of returns to the quasi-fixed or primary factors, and constrained by temporary equilibrium forces, from optimally adjusting to their long-run level.

---

[9]It is assumed that both country's prices are measured in the same currency, achieved by conversion through the exchange rate.

$MIC^t_{i,j}$ is a measure of changes in relative competitiveness between countries i and j, at any point in time t, with reference to the base period t=0. If in the base period $MIC^0_{i,j}=1$, then the industry in country i is argued to be relatively more competitive than the industry in country j, if $MIC^t_{i,j} > MIC^0_{i,j}$, where t and 0 are time superscripts denoting the tth and base period respectively. If with reference to the base period countries i and j possessed no competitive edge over each other, given the relative competitive advantage possessed by country i in time period t, a gap in competitiveness between the two countries will now exist. This will be referred to as a 'competitiveness gap' in favor of country i. Relative competitive advantage\disadvantage in subsequent time periods by country i will cause a widening\narrowing of the competitiveness gap with country j. Conversely, if $MIC^t_{i,j} < MIC^0_{i,j}$, then beginning from a position of 'balanced competitiveness' ($MIC^0_{i,j}=1$), between the two countries in the base period t=0, the industry/sector in country i is argued to be less competitive than the industry in country j, and a competitiveness gap is said to exist in favor of country j. When $MIC^t_{i,j} = MIC^0_{i,j}$, neither country possesses a competitive advantage in this sector/industry, so there is no change in competitive positions and no competitive gap exists between the two countries.

The difference in restricted profit between country i and j, at a given point in time, is found by differenting the restricted profit function given in equation (1), with respect to D_c. In the interest of brevity the derivation of the measures will be by-passed and we proceed instead directly to the presentation of the results.

After differentiating equation (1), and making the appropriate substitutions the short-run MIC is given by:

$$\frac{dln\pi_{it}}{dD_c} = [(1+\epsilon_{sit})\Re_{it} - (\epsilon_{C_{it}\,Y_{dt}} \cdot \epsilon_{sit})\mathfrak{C}_{it}]\ \frac{dlnP_{sit}}{dlnD_c} +$$

$$[(\epsilon_{Y_{dt}\,w_{mdt}}\Re_{it} + S_{X_{it}}) - (\epsilon_{C_{it}\,Y_{dt}} \cdot \epsilon_{Y_{dt}\,w_{mdt}})\mathfrak{C}_{it})] \cdot \frac{dlnw_{it}}{dlnD_c} + \Re_{it}\frac{\dot{A}}{A}\,\frac{dT_t}{dD_c} \qquad (2)$$

The ratio of costs to restricted profit, and the ratio of total revenue to restricted profit are represented by: $\frac{C_{it}}{\pi_{it}} = \mathfrak{C}_{it}$ and $\frac{P_{sit} \cdot Y_{sit}}{\pi_{it}} = \Re_{it}$, respectively.

The cost and supply elasticities characterize the short-run consequences of changes in the exogenous variables, conditional on the level of z_{it}. $\epsilon_{Y_{it}, w_{it}}$ is the output consequence arising from a change in variable or primary input prices. The elasticity of supply $\epsilon_{S_{dt}}$ represents the change in output in the short-run when an exogenous change in the product price occurs. Under the assumption of the restricted technology as an appropriate characterization of production interrelationships, all the cost elasticities are dependent on the extent of quasifixity and capacity utilization. The elasticity of cost with respect to output $\epsilon_{C_{it}, Y_{dt}}$ is the change in total costs resulting from a one unit change in output.

The MIC measure given by equation (2), represents the fundamental equation of measured differentials in short-run industry\sectoral competitiveness over time.

The first two terms of equation (2), represent the restricted terms of trade effect. The term 'restricted' is used to qualify the use of the terms of trade concept, since it is not adjusted for changes in either traded inputs or for intermediate products traded between countries.

The second set of terms highlights the impact of relative factor price movements on MIC. The hypothesis implied by the third set of terms in equation (2), is that the sector with the higher rate of technological progress, relative to the sector in the other country, will have a slower rate of growth in costs, which will result in higher rates of profit. Owing to the ambiguity in both the sign and magnitude of the final effects, little else can be advanced barring empirical estimation of the model.

V. Bibliography

Ahearn, M., D. Culver, and R. Schoney. "Usefulness and Limitations of COP Estimates for Evaluating International Competitiveness: A Comparison of Canadian and U.S. Wheat." American Journal of Agricultural Economics 72(5)(1990): 1283-1291.

Balassa, B. "Recent Developments in the Competitiveness of American Industry and Prospects for the Future," in Factors Affecting the U.S. Balance of Payments, Subcommittee on International Exchange and Payments, Joint Committee of the Congress, 1962.

Brinkman, G. "The Competitive Position of Canadian Agriculture." Canadian Journal of Agricultural Economics 35(3)(1987):263-288.

Dogramaci, A. "Perspectives on Productivity." Productivity Analysis: A Range of Perspectives., ed. A. Dogramaci. Boston: Martinus Nijhoff Publishing, 1981.

Griliches, Z. "Research Expenditures, Education and the Aggregate Agricultural Production Function." American Economic Review, 54(1964):961-974.

Harriss,L. "Basic Forces in Productivity Growth." In Productivity Analysis: A Range of Perspectives., ed. A. Dogramaci. Boston: Martinus Nijhoff Publishing, 1981.

Hicks, John. Essays in World Economics. Oxford: Clarendon Press, 1959.

Islam, S. "Currency Misalignment: the Case of the Dollar and the Yen," Quarterly Review 8(4)(1983):49-60.

Johnson, Harry. Economic Development and International Trade. Cambridge, 1968.

McCulloch, Rachel in Baldwin, Robert and Richardson J. David. International Trade and Finance Readings. Little Brown and Company. Boston, 1986.

Morici, P. The Global Competitive Struggle: Challenges to the United States and Canada. Canadian-American Committee, Washington, D.C., 1984.

Morici, P. Meeting the Competitive Challenge: Canada and the United States in the Global Economy. Canadian-American Committee, Washington, D.C., 1988.

Norton, G., and J. Davis. "Evaluating Returns to Agricultural Research: A Review." American Journal of Agricultural Economics 63(1981):685-699.

Porter, M.E. The Competitive Advantage of Nations. New York: The Free Press Inc., 1990.

Ruttan, V. "Agricultural Research and the Future of American Agriculture." In the Future of America as a Strategic Resource eds. S.S. Batie and R. Healy, Washington, D.C.: The Conservation Foundation, 1980.

Shumway, C. "Models and Methods Used to Allocate Resources in Agricultural Research: A Critical Review." In Resource Allocation and Productivity in Natural and International Agricultural Research, ed. T.M. Arndt, D. Dalrympia, and V.W. Ruttan, Minneapolis: University of Minnesota Press, 1977.

Thompson, R. A Survey of Recent U.S. Developments in International Agricultural Trade Models. Washington, D.C.: USDA, ERS, International Economics Division, 1981.

Vollrath, Thomas. "Revealed Competitiveness for Wheat." ERS Staff Report No. AGES861036. USDA, Economic Research Service, (1987).

Warley, T. Agriculture in an Interdependent World: U.S. and Canadian Perspectives. Canadian-American Committee, Washington, D.C., 1977.

A. Budget:

LINE ITEM	COST[1]
Stipend-1/3 Ph.D.	$13,600
Computer-time and software[2]	$2,200
Incidental Supplies	$500
Publication Costs	$200
Purchase of Data	$900
TOTAL COST	$17,400

Notes:

1. All currency denominated in U.S. Dollars.
2. Based on 11 man-months of access time, including software. Essentially two software packages are required, an econometrics package and a word-processing package.

Index